Adam Johnstone

BIOLOGY
FACTS & PRACTICE FOR A LEVEL

OXFORD

UNIVERSITY PRESS

OXFORD
UNIVERSITY PRESS

Great Clarendon Street, Oxford OX2 6DP

Oxford University Press is a department of the University of Oxford.
It furthers the University's objective of excellence in research, scholarship, and education by publishing worldwide in

Oxford   New York

Athens   Auckland   Bangkok   Bogotá   Buenos Aires   Cape Town   Chennai
Dar es Salaam   Delhi   Florence   Hong Kong   Istanbul   Karachi   Kolkata
Kuala Lumpur   Shanghai   Madrid   Melbourne   Mexico City   Mumbai
Nairobi   Paris   São Paulo   Shanghai   Singapore   Taipei   Tokyo   Toronto
Warsaw

with associated companies in Berlin   Ibadan

Oxford is a registered trade mark of Oxford University Press in the UK and in certain other countries

British Library Cataloguing in Publication Data

Data available

ISBN 0 19 914 766 3

Typeset by Magnet Harlequin

Printed and bound in Great Britain

Author's acknowledgements

Thanks to all the students, over the years, who didn't get it first time round and who made me explain it again in a different way. Thanks also to Jaimie, for the endless loan of his office and for his help with the concept tests.

INTRODUCTION

There are a large number of revision guides on the market at the moment which attempt to provide a concise summary of all the material in the new AS and A2 specifications. This is not one of those guides. This is a specification support book. It provides a clear and easily accessible introduction to 28 key topics that occur in all the current specifications, with an emphasis on conceptual understanding rather than on total coverage.

This book also breaks away from the standard revision guide format in providing a simple and linear explanation of each of the 28 key topics, as opposed to a deluge of information and graphics. It is my conviction, after over a decade of testing material out on students, that a simple, linear approach is the best way to get information across. Telling a story is what it's all about.

As well as the topic explanations, this book contains a resource collection of 28 recall tests and 28 concept tests. There is a recall test and a concept test after every explanation spread. Answer lines are included, in exam style format, so that students can write in the book. The recall tests should allow self testing, or class testing. The concept tests, which have been designed to include a variety of question types, should aid with the development of exam technique as well as providing opportunities for class discussion.

So here it is. A conceptually focused revision guide that targets key topics in a linear, storytelling style and backs this up with lots of questions. I hope you find it useful.

ADAM JOHNSTONE

Adam Johnstone teaches Biology at d'Overbroeck's, an independent school in Oxford.

CONTENTS

BIOLOGY A LEVEL

What topics are covered in a biology A level?

Biology is the study of life itself. In a biology A level you will learn what it is that makes life tick. Some of the things you study will be sub-microscopic, like the tiny protein pumps in the membrane of a living cell, a hundred thousand of which can be fitted into one millimetre. Some of the things you study will be vast, like the global ecosystem itself, shuddering under the impact of human civilization. Some topics will seem very abstract and difficult to connect to, like the biochemistry of photosynthesis or the flow of energy through a food web. Some topics will seem very close to home. Why do we sweat or fart or cry? Why do we bleed when cut or turn white when we are frightened? In a biology A level you will get a brief glimpse of some of the amazing design solutions that animals and plants have come up with in their struggle to survive. Skins that change colour. Eyes that can see in infrared. Kidneys so effective that their owners need never drink. Above all else, you will learn about living things; things that hunt and mate and replicate, things that can hold their breath for an hour, things that can change their sex, things with penises ten times the length of their own body. There is nothing as strange as the reality of the living world. There may be a lot to learn in a biology A Level, but it's all interesting.

How hard is a biology A level?

Biology is a science subject. Though it doesn't contain as many mathematical and abstract topics as physics or chemistry, it does have some maths in it and it does contain topics that require a technical turn of mind. For this reason, students whose talents lie in the discursive and literary world of the humanities can find it challenging. Equally, students who are mathematically inclined and who enjoy the clear logic of physics and chemistry can also find biology a challenge. This is because the living world does not follow simple equations. Life is not logical, but partial, contradictory, and full of exceptions. A biologist must have the ability to think laterally and the flexibility to spot patterns in a mass of apparently random data. Finally, a biology A level has a large practical component. It is not an A level for purely abstract thinkers who are afraid to engage with the real world. You are going to have to handle test tubes at some point if you choose to do biology. You are going to have to learn the skills that are needed to design and then run an actual experiment. Have I put you off yet?

What other subjects go well with biology?

Chemistry is traditionally supposed to be the ideal complement to biology. This is less true now than it used to be. While studying chemistry will certainly help you to understand the biochemical component of a biology A level, it isn't an absolute requirement. Most of the basic chemistry that you need to know is covered in the first unit of this book. Maths is always useful, especially statistics, but, again, it is not a requirement. Geography, which is less often mentioned, is also a good complement to biology. Geographers will find that the ecology component of the biology A level overlaps significantly with their study of physical geography. Probably the tightest link is between biology and sports studies or PE. Much of the material on nerves, muscle, the heart and the lungs is shared between these subjects.

What is the AS/A2 system?

In the old days, most students did three different A levels. Each A level involved a two-year course followed by a set of exams. Students would obtain a single grade, from A to U, for each A level completed. When applying to university, their three grades would be combined together to give a UCAS point score, varying from 30 to 0. Recently, with the introduction of the AS/A2 system, the traditional two-year A level course has been split into two parts. After one year of studying a subject, students take an AS exam. At this point they can stop, if they want, or they can complete a second year of study and then sit an A2 exam as well. If they do both an AS and an A2 exam then they achieve a single A level grade in that subject and score an appropriate number of UCAS points. If they just do the AS, they achieve a grade which is worth exactly half the value of an equivalent A level grade. In most cases, a typical student is likely to do four or five ASs in their first year of study and then take three of these on to A2, though this may vary from school to school.

What are key skills?

Key skills have been part of the GNVQ qualification for quite a while but they have only recently been added to the A level system. The aim of this addition is to guarantee that sixth form students have basic skills in numeracy, information technology, and communication. Though they are not making it compulsory, the government hope that sixth form students in most schools will work to obtain a level 3 key skills qualification in each of these three areas. In order to achieve such a qualification, students must accumulate written or other evidence that they have the appropriate skill and also sit and pass a brief exam. It is the accumulation of evidence that requires the real work when you are going for a key skills qualification, and this is where A level subjects fit in. Your teachers may organize things so you are able to use the work that you're doing as part of your regular A levels as evidence towards your key skills qualifications. This means that you won't have to do a lot of extra work especially for key skills. If you do a presentation to the rest of the biology class, you could use this as evidence for the communication key skill. If you have plotted graphs, based on the data that you collected on a field trip, you could use these as evidence for the numeracy key skill, etc.

What use is a biology A level?

If you want to be a doctor, a dentist, a physiotherapist, or a vet, then you definitely need to study biology A level and you will probably need to get a grade 'A'. All of these professions require a detailed knowledge of human or animal anatomy. Nurses, opticians, and pharmacologists are also required to know their human biology. If you want to do a university degree in one of the many specialist biology subjects, whether it's marine biology or neuroscience, or you just want to do a straight biology course, then an A level in Biology is an obvious must. There are all sorts of challenging biology-related jobs opening up these days, from genetic engineering to environmental consultancy. Finally, even if you aren't going to do a job that is related to biology, it is still a subject worth studying. It will give you a better appreciation of the incredible mechanism that is your own body and of the extraordinary diversity of living things. It will allow you to comment, in an informed way, on the crucial issues of our age, from the war against disease to the advance of GM foods to the sad destruction of the global environment. If you aren't interested in life itself, then what are you interested in?

EXAM TECHNIQUE

These days, with the introduction of modular papers and the AS/A2 system, A level students are having to sit more exams than ever before. Back in the 1990s, if you'd been doing a biology A level, you would only have had to sit a couple of papers at the end of the whole two-year course. Today, most students have to sit exams in the January of the AS year, only one term after finishing GCSEs. Over the whole AS/A2 course, there are likely to be at least four exam sittings. Modern exams are also trickier than they used to be. You may have heard that A levels today are easier than they once were, but this is a bit misleading. While it is true that they require you to learn less, they make up for this by requiring a lot of analytical skill. A modern biology exam is full of passages that have to be read and understood and graphs that need interpreting. Given the number of exams that you will have to do, and the level of analytical ability that they will require, it is obviously important that you start developing your exam technique as soon as possible.

The best way, by far, to develop exam technique is to get in a lot of practice. Your school will probably organise mock exams for you. Do take these seriously! If you don't revise for a mock exam and then try your hardest, you won't be able to learn from it properly. As well as mock exams there are also past papers. Get hold of as many of these as possible and see if you can locate a kind teacher to mark them for you when you have had a go at them.

The questions in this book should also prove helpful, as long as you actually do them. For more advice on how to use the question sections in this book as part of your revision process, take a look at the section titled 'How to use this book', on page x.

Finally, there are few pieces of advice which you should pay attention to:

Read the question

This may seem incredibly obvious but it is amazing how even good students, who have revised hard and know a lot of biology, can still manage to lose marks because they don't read the questions carefully enough. Doing an exam, even if you have revised for it, is a stressful process and it is very easy to get panicked or carried away so that you just start scribbling down everything that you know after one glance at the question. Don't do this. Put your pen down and pause before each new question. Look at it, read it all the way through and think about it for a couple of minutes before you pick your pen up again. Believe me, this doesn't cost much in time and it really does help. Part 'c' of the question may give you a big clue to the answer required in part 'a'. After a bit of time spent puzzling things out, you may realize that part 'b' is getting at something completely different to what you originally imagined. In other words, think before you write.

Answer the question

This means the question in front of you: not an imaginary question a bit like it, not a similar question that you have done before, but this question. Don't just churn out everything you know about the topic. Think! Work out what the examiners are getting at. What points are likely to be credited on the mark scheme? What facts are relevant? If you are asked to link A to B, then specifically link them. If it you are asked to explain something, making reference to a diagram, then make sure you refer to the diagram.

Style

In a biology exam you are not being tested on your ability to produce stunning English prose. Far too many students go wrong by attempting long complex sentences, full of sub-clauses, that wander in and out of the various tenses and voices possible in English. 'The enzyme has an active site that is on its surface so fits the substrate and makes a reaction because it was the right shape so that it is specific.' That kind of thing. It is much easier and more effective to write a series of short, clear sentences. If it helps, you can think of these as bullet points placed end to end. 'Each type of enzyme has an active site. This is a region on the enzyme's surface with a particular shape. Only substrates with a particular shape will fit into this active site. Because of this enzymes are described as substrate specific'. This may sound terrible, and rather robotic, but it is easier to produce, easier to understand, and it will get the marks.

Length

All biology exams, these days, require you to write your answers on the exam script and provide you with a number of dotted lines to write on. Though it is difficult to get a straight answer out of them, most of the awarding bodies claim that they do not directly penalize answers that carry on beyond the allotted number of dotted lines. So, if you are really desperate, you can add in a couple of additional sentences. It is worth remembering, however, that you are not being marked on how much you write. If this were the case, it would be the people with the smallest writing who would get all the 'A' grades. The number of lines provided tells you how much the examiners expect you to write in order to get full marks. If three lines are provided that means that the maximum mark answer can be written in three lines. If you feel you can't fit your answer in, then you've probably got the wrong answer. Equally, if your answer's only one line long, then you've probably missing something out.

Confidence

However well you have revised, you are bound to encounter a lot of material in a modern biology A level exam that you have never come across before. Do not be thrown by this. It is not a problem. Because modern exams are analytical, the examiners fill them, on purpose, with scenarios and contexts that they don't expect students to be familiar with. They are not doing this because they are evil but because they want to see whether you can apply what you know to a slightly unusual situation. 'Describe how glucose diffusion across the membrane of an armadillo pancreas cell will vary with temperature.' It doesn't matter, in this case, that it's an armadillo or that it's a pancreas. It doesn't even matter that it's glucose. This is simply a question about diffusion (see unit 6). If you have confidence in yourself and your knowledge you should be able to find your way through all the irrelevant information and unusual diagrams to the simple biology that lies beneath.

Timing

This is obvious. Just make sure that you give yourself enough time to attempt all the questions. There is no point spending 15 minutes on the last part of question 4, which is only worth a couple of marks, if this prevents you from answering the whole of question 5. Exam practice should give you a feel for the appropriate speed to work at. If you consistently finish early, this means that you aren't spending enough time on reading the questions and planning your answers. If you consistently run out of time, you may need to devise ways of working faster.

HOW TO USE THIS BOOK

This book is not a revision guide in the usual sense, since it is not designed to provide a complete and abbreviated summary of everything in the AS and A2 specifications. Nor is it a textbook, because it doesn't use complex language or go into massive technical detail. Instead, it is designed as a learning support book. The 28 units that it contains provide clear and easily accessible introductory explanations to 28 key topics that occur in all the biology specifications. The questions that follow each unit provide a basis for self-testing, so that you can check what you have learned, and an opportunity to stretch your analytical skills.

The units

Probably the best time to read a unit in this book is just after you have been taught the relevant topic at school. You should find the style of explanation easy to get to grips with and more accessible than that in most text books and by backing up what you have been taught in class you should achieve a clearer understanding of the topic. Understanding is what this book is all about, because understanding is the key to success in the modern, analytical AS/A2 system. Don't expect the units in this book to cover every single fact that you will need to know. If you want a dry and all-inclusive list, then take a look at the specification or at an official specification revision guide.

The recall tests

The best time to do a recall test is immediately after you have read a unit. This will only work, of course, if you take the test seriously and try doing it all before you look up the answers. Set yourself a challenge before you even start reading the unit. See if you can read it and then get 10 out of 10 on the test. If you take this approach, you will find that you end up absorbing the material in the unit a lot more effectively.

The concept tests

The concept tests are something you might want to spend a bit more time mulling over. All of the questions in a concept test can be answered using information covered in the unit, so they don't require you to look in any other books, but many of them do require some serious lateral thinking. If you find yourself completing the concept tests easily, then you are probably a natural biologist. If you're finding them hard, don't be worried. Spend a couple of days thinking about them, if you want, before you look up the answers. You may well find that you were getting close to an answer but just lacked the confidence in yourself. It is through doing the concept tests that you will really develop the mindset needed to take on the analytical questions in the A level.

Exam board specifications and modules

The table below shows where the units in this book complement the different modules of the various A Level specifications. This should help you to link what you are doing in school with the material in the book.

	AS			A2		
	Mod 1	**Mod 2**		**Mod 5**	**Mod 6**	
AQA Biology specification A	1–7, 18–20	10–13, 23		8, 9, 14–17, 25–28	21, 22	
	Mod 1	**Mod 3**		**Mod 5**	**Mod 7**	
AQA Human Biology specification A	1–7, 18–20	10–13		8, 9, 14–17, 25–28	22, 23	
	Mod 1	**Mod 2**	**Mod 3**	**Mod 4**	**Mod 5**	**Mod 8**
AQA Biology specification B	1–7, 18, 22	10–14	19–21	8, 9, 15–17	25–28	23
	Mod 1	**Mod 2B**	**Mod 3**	**Mod 4**	**Mod 5B**	
Edexcel Biology	1–7, 10–11, 13	14, 18–24	25–28	8	9, 12, 15–17	
	Mod 1	**Mod 2H**	**Mod 3**	**Mod 4**	**Mod 5H**	
Edexcel Human Biology	1–7, 10, 11, 13	14, 18–20, 22, 23	25–28	8	12, 15, 16	
	Mod 1	**Mod 2**	**Mod 3**	**Mod 4**	**Mod 5**	
OCR Biology	1–7, 10, 11, 13, 14, 25–27	18	19–21	8, 9, 14, 16, 17	12, 22–24, 28	

AN INTRODUCTION TO MOLECULES

● Living organisms are built out of molecules: carbohydrates and lipids, catalytic proteins, information-storing molecules of DNA. It is the interaction of all these molecules that makes life possible. You don't have to be doing a chemistry A level in order to understand how molecules work, but there are some basic bits of chemistry which are worth knowing.

● A **molecule** is a collection of **atoms**, which are joined together by **bonds**.

● There are many different types of atom and each type has different properties. An **element** is a substance that contains only one type of atom. The atoms that make up the element carbon are called 'carbon atoms', those that make up the element oxygen are called 'oxygen atoms', etc.

● A **molecular formula** tells us what type of atoms are found in a molecule and how many there are of each type. The molecular formula of water is H_2O. This tells us that it contains two hydrogen atoms and one oxygen atom. The molecular formula of α-glucose is $C_6H_{12}O_6$. This tells us that it contains six atoms of carbon, twelve of hydrogen, and six of oxygen.

● A **structural formula** is a diagram which shows the actual arrangement of the atoms in a molecule. The structural formulae of water and of α-glucose are given below.

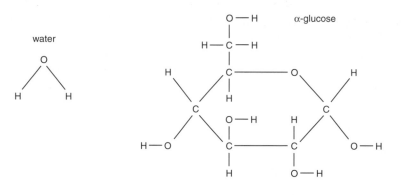

● In the diagram, you can see that there are lines joining the atoms together. These lines represent **covalent bonds**. This type of bond is the one most commonly used to link atoms together in the construction of biological molecules.

● Each type of atom makes a different number of covalent bonds with its neighbours. From the structural formulae, you can see that carbon atoms make four bonds, oxygen atoms make two, and hydrogen atoms make one. The number of covalent bonds that an atom makes with its neighbours is known as its **valency**. It is helpful to know the valency of different atoms when you are trying to remember a structural formula.

● In the days before molecules were really understood, it was thought that living organisms must be made out of a completely different sort of material to non-living things. How else could they move and grow and reproduce? Today, we know that both living and non-living things are composed of the same atoms. The important difference lies in the way that these atoms are combined together to make molecules. The air, a non-living substance, contains molecules of nitrogen (N_2), oxygen (O_2), carbon dioxide (CO_2), and water (H_2O). As you can see, all these molecules are pretty small and simple. They are **inorganic molecules**. The molecules used to build a living organism are much larger and much more complex. These **organic molecules** may contain thousands of atoms of carbon, hydrogen, oxygen, and nitrogen.

● All organic molecules are based on rings or chains of carbon atoms. Carbon is used because it has a valency of four, which means that you can stick carbon atoms together in a chain, or a ring, and still have spare bonds to attach to other atoms or groups of atoms. The only other commonly occurring atom with a valency as high as that of carbon is silicon. Sci-fi writers may imagine organisms built from silicon, but all the life on Earth that we are aware of is carbon based.

● All of the really big organic molecules are **polymeric**. This means that they are made by sticking many smaller organic molecules together. These smaller molecules are known as **monomers** and they are joined together by **condensation** reactions.

● A condensation reaction involves the removal of an OH group from one monomer, and an H from another, to produce a covalent bond between the two of them. A molecule of H_2O is produced in the process. The opposite of condensation is **hydrolysis**, where H_2O is added in order to split two monomers apart.

● The different types of monomer, the names of the covalent bonds that link them, and the polymers that they form are listed to the right.

● Covalent bonds may be used to link monomers into a polymer, but the ultimate shape of most macromolecules (molecules containing a very large number of atoms) is dependent on a weaker, though equally important, type of bond called the **hydrogen bond**.

● Atoms of oxygen and nitrogen are described by chemists as **electronegative**. What this means, in practice, is that whenever an oxygen or a nitrogen atom is covalently linked to a hydrogen atom, the hydrogen atom will acquire a small positive charge. This positively charged hydrogen can form a link, called a hydrogen bond, with other electronegative atoms that are nearby.

Monosaccharides are joined by **glycosidic bonds** to make **polysaccharides** (see unit 2).

Amino acids are joined by **peptide bonds** to make **proteins** (see unit 3).

Nucleotides are joined by **phosphoester bonds** to make **polynucleotides** (see unit 10).

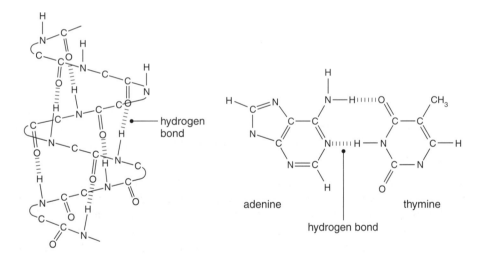

Hydrogen bonding in a coiled polypeptide chain, left (see unit 3), and hydrogen bonding between two organic bases, right (see unit 10)

● In macromolecules, such as proteins, where the polymeric chain is folded back on itself, it is hydrogen bonds that are used to maintain this folding, by linking one bit of the chain with another (see unit 3). Hydrogen bonds can also be used to link two polymeric chains together, as seen in cellulose (unit 2) and DNA (unit 10).

● If an organic molecule is charged, or **polar**, it can form hydrogen bonds with surrounding molecules of water. Small polar molecules, that can link with water in this way, are described as **water soluble**. If you add them to water, they dissolve. Examples include amino acids, monosaccharides, and nucleotides. All of these contain OH or NH groups, which is what gives them their charge and allows the hydrogen bonding with water to occur. Since most chemical reactions will only happen if the molecules involved are dissolved in water, you can see how important hydrogen bonding is.

TESTS

RECALL TEST

1 What is a molecule?

_____

_____ (1)

2 What is an element?

_____

_____ (1)

3 What is the valency of carbon?

_____ (1)

4 Give the molecular formulae of four different inorganic molecules that are found in the air.

_____

_____ (1)

5 What atoms are removed during a condensation reaction?

_____ (1)

6 What type of covalent bond is used to link two monosaccharides together?

_____ (1)

7 What would complete hydrolysis of a protein produce?

_____ (1)

8 Name two electronegative atoms.

_____ (1)

9 Which is stronger, a hydrogen bond or a covalent bond?

_____ (1)

10 Name two types of macromolecule in which separate polymeric chains are linked together by hydrogen bonds.

_____ (1)

(Total 10 marks)

CONCEPT TEST

The structural formulae of three different molecules are shown below.

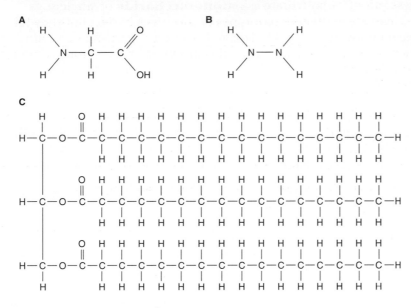

1 What is the molecular formula of molecule **B**?

_____ (1)

2 What is the valency of nitrogen? How did you deduce this?

_____

_____ (2)

3 What information about the structure of molecules **A**, **B**, and **C** is *not* provided by their structural formulae?

_____

_____ (1)

4 Which of the molecules shown are organic molecules? Explain your reasoning.

_____

_____ (2)

5 Of the two molecules **A** and **C**, which is water-soluble and which isn't? Explain your reasoning.

_____

_____

_____

_____

_____ (4)

(Total 10 marks)

CARBOHYDRATES AND LIPIDS

- **Carbohydrates** are organic molecules containing the elements carbon, hydrogen, and oxygen. They are found in all living organisms.

- The simplest sort of carbohydrate is a **monosaccharide** or single sugar molecule. There are many different types of monosaccharide, but they all have the general formula $C_nH_{2n}O_n$. This tells us that they always contain twice as many hydrogen atoms as they do carbon or oxygen atoms. Different types of monosaccharide are divided up according to the exact number of carbon atoms that they contain.

- **Triose** monosaccharides have three carbon atoms.

- **Pentose** monosaccharides have five carbon atoms. Examples include **ribose** and **deoxyribose**, which are used in the construction of larger molecules like RNA (ribonucleic acid) and DNA (deoxyribonucleic acid) (see unit 10).

- **Hexose** monosaccharides have six carbon atoms. Examples include **glucose**, **galactose**, and **fructose**, which all have the formula $C_6H_{12}O_6$. Because they are built out of the same atoms, arranged differently, we describe these hexose sugars as **isomers**. The structural formula of **α-glucose** is shown below.

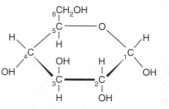

As you can see in the diagram, the carbon atoms are numbered, clockwise, from one to six. An alternative form of glucose is **β-glucose**, in which the OH group, attached to carbon atom number one, sticks up above the ring, with the H below.

- Two monosaccharides can be joined together by a **condensation reaction**, in which water is removed, to produce a **disaccharide**. The bond formed between the two monosaccharides is known as a **glycosidic bond**. A diagram showing the formation of the disaccharide **maltose**, from two molecules of α-glucose, is shown below.

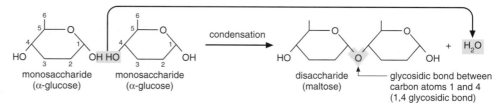

The bond formed in this case is called a **1,4 glycosidic bond**, because it links the first carbon of one monosaccharide with the fourth carbon of the other.

- Other disaccharides include **lactose** and **sucrose**. Here is a summary of disaccharide formation.

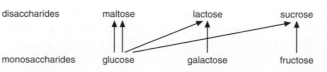

- All the monosaccharides and disaccharides are sugars. They come in the form of sweet, white crystals that can dissolve in water. The sugar that we put in our tea and coffee is sucrose. This is obtained from plants, such as sugar cane, where it is naturally used as a transport carbohydrate (see unit 21).

- If many monosaccharides are joined in a chain the result is a **polysaccharide**. Examples include **glycogen**, **starch**, and **cellulose**.

- Glycogen and starch are **storage carbohydrates**. They are manufactured when an organism has spare glucose to hand, and broken down again when glucose is running low. Glycogen is used by animals and starch by plants.

- Both molecules are built from coiled chains of α-glucose. Glycogen consists of branching chains, with **1,6 glycosidic bonds** at the branch points. Starch is a mixture of two types of molecule: **amylopectin**, which is branched, and **amylose**, which isn't.

- Because of their large size, both glycogen and starch are **insoluble**. This means that they can't diffuse out of the cells they're stored in, or cause osmotic problems. Because they contain branching chains, they are **compact**, which means that they can store a lot of energy in a small space.

- Cellulose is a **structural carbohydrate** that is used to build **cell walls** in a plant (see unit 5). It consists of straight chains of β-glucose which are linked together by hydrogen bonds to form a rigid structure, similar to scaffolding.

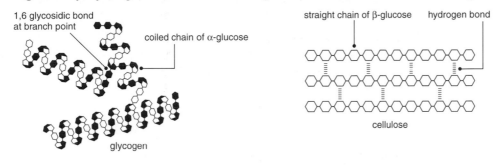

Glycogen and cellulose

- **Lipids**, like carbohydrates, are built out of the elements carbon, hydrogen, and oxygen. Despite this, they have very different properties.

- The most common types of lipid are fats and oils, or **triglycerides** as they are officially known. A triglyceride is constructed from a **glycerol** head which is joined by condensation reactions to three **fatty-acid** tails, as shown below.

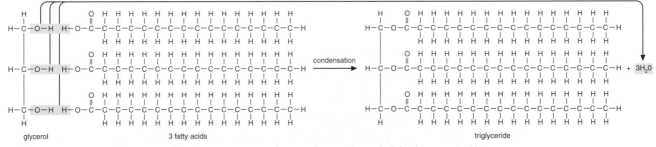

- Because of a lack of OH and NH groups, triglycerides are insoluble (see unit 1). This means that, like starch and glycogen, they make ideal **energy-storage** molecules. Fats tend to be used as a long-term energy store, which is only accessed after all of an organism's carboyhdrate reserves have been used up. Triglycerides, in the form of body fat, are also used to provide insulation and protection for delicate organs.

- Fatty acids, which are used in the construction of a triglyceride, come in two varieties, **saturated** and **unsaturated**. You have probably heard about these in relation to diet. The fatty acids shown in the triglyceride diagram above are saturated, as found in most animal fat. Unsaturated fatty acids have **double bonds** between carbons at key points, which causes the fatty-acid tail to kink or bend. They are most commonly found in fish and plant oils, and are less likely to contribute to heart disease.

- **Phospholipids** are similar in structure to triglycerides, except for the fact that they have only two fatty-acid tails and a charged phosphate group attached to their glycerol head. The charged head and uncharged tails of a phospholipid molecule lead to some unusual properties. It is these properties that allow phospholipids to function as a component of the cell membrane (see unit 6).

- Other types of lipid include **waxes**, such as those that coat the surface of a leaf, and **steroids**, like the sex hormone **testosterone**.

TESTS

RECALL TEST

1 What is the molecular formula of a triose monosaccharide?

_____ (1)

2 Draw out the structural formula of β-glucose.

(1)

3 What would be produced by the hydrolysis of sucrose?

_____ (1)

4 List three properties common to all monosaccharides and disaccharides.

_____ (1)

5 Name two structural features of starch which make it a good energy-storage molecule.

_____ (1)

6 Name a polysaccharide which is made from β-glucose.

_____ (1)

7 What is the structure of a triglyceride?

_____

_____ (1)

8 Why are fats and oils insoluble?

_____

_____ (1)

9 What is the difference between saturated and unsaturated fatty acids?

_____

_____ (1)

10 List five functions of lipids.

_____

_____

_____ (1)

(Total 10 marks)

CONCEPT TEST

1 A key which can be used to identify five different types of carbohydrate is shown below:

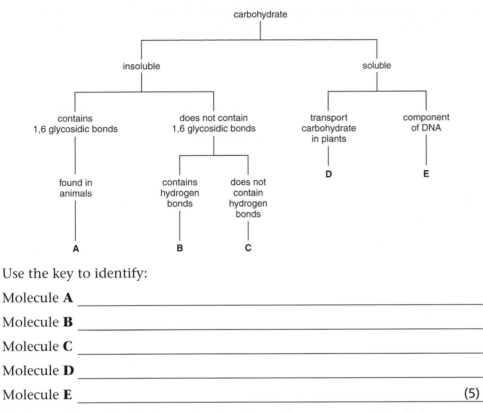

Use the key to identify:

Molecule **A** _____

Molecule **B** _____

Molecule **C** _____

Molecule **D** _____

Molecule **E** _____ (5)

2 Work out the molecular formula of maltose. Show your working below.

(2)

3 Both lipids and carbohydrates can be used as energy-storage molecules. Lipids store more energy per gram but carbohydrates are easier to break down. Use this information to explain why animals use both glycogen and fat as energy stores while plants only use starch.

_____

_____

_____

_____

_____

_____ (3)

(Total 10 marks)

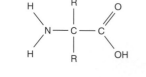

Unit 3 PROTEINS

- Proteins are large, complicated molecules containing the elements carbon, hydrogen, oxygen, nitrogen, and, in most cases, sulphur. They are found in all living organisms.

- Proteins are built out of **amino acids**. These are small molecules which are stuck together in a long chain to make a **polypeptide**. This polypeptide chain is then folded and coiled and, in some cases, attached to other polypeptide chains to produce the final protein.

- The general formula of an amino acid is shown left.

An amino acid

- There are twenty different types of amino acids which can be used in building a protein. They all have the same basic structure but have different collections of atoms in the place labelled R on the diagram. We say that different amino acids have different **R groups**.

- Two amino acids can be joined together using a condensation reaction in which an H and an OH are removed to release a molecule of water. In this way a **peptide bond** is formed between the two amino acids, as shown below.

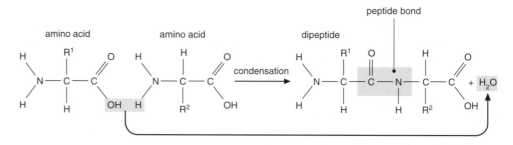

If two amino acids are stuck together like this the result is a **dipeptide**. If many amino acids are stuck together the result is the polypeptide that we mentioned earlier.

- The **primary structure** of a protein is the order in which the amino acids are linked together to make the polypeptide chain.

- The **secondary structure** of a protein is the way that the polypeptide chain is coiled or folded. There are two very common ways in which this can happen. Either the chain is coiled like a spring to produce an **α-helix** shape, or lengths of the chain can line up side by side to produce a **β-pleated** sheet.

An α-helix (left) and a β-pleated sheet (right)

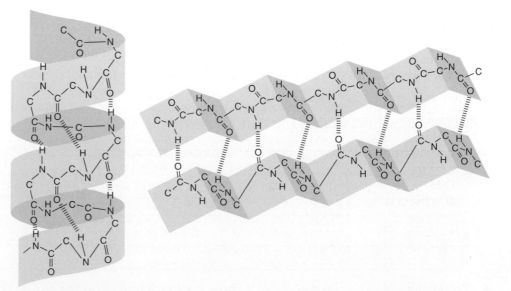

The dashed lines (IIIIII) in the diagram represent hydrogen bonds which hold the coiled chain in its new shape.

10

- Having secondary structure is as complicated as some proteins get, but many large globular proteins, like enzymes, may have **tertiary structure**. This is the coiling and folding of the already coiled and folded chain, as below.

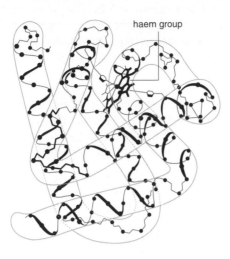

haem group

The coiled black line represents the polypeptide chain and each black dot is an amino acid. You can see that the chain has been folded to make an α helix and that this coiled helix has then been further folded to produce a roughly globular shape. (The **haem group** is a non-protein molecule that has been added in as well.)

- The tertiary coiling is held in place by hydrogen bonds, as was the secondary structure, but there are other, stronger bonds involved as well, including **disulphide bonds** and **electrovalent bonds**.

- Some proteins even have **quaternary structure** in which several different polypeptide chains are attached to each other. In **haemoglobin**, the oxygen-carrying molecule found in red blood cells, there are four chains. Each chain is folded into a globular shape and then all four are stuck together.

- The really important thing about a protein is the order in which the amino acids are linked to make the polypeptide. This is controlled by **DNA** (deoxyribonucleic acid) during **protein synthesis** (see unit 11).

- ***Which amino acids are stuck together, and in which order, determines how the polypeptide chain will fold up, and this determines the shape of the protein.***

- Because there are so many ways of ordering the twenty different amino acids, proteins can be built in almost any shape. As a result, living organisms can use proteins to do a whole range of different jobs. Each type of protein has a shape designed to suit its particular function.

- Here is a table listing some proteins and their functions.

Protein	Description
haemoglobin	a protein that carries oxygen (see unit 20)
amylase	an enzyme that breaks down starch (see unit 22)
FSH	a hormone involved in the menstrual cycle (see unit 23)
fibrinogen	a plasma protein involved in blood clotting
myosin	a contractile protein found in muscle
keratin	a fibrous protein found in hair and nails
albumin	an energy-storage protein found in egg white

- Proteins are also used to make bacterial flagella (see unit 5), active transport pumps in the plasma membrane (see unit 6), ribosomes (see unit 11), and spindle fibres (see unit 13). On almost every page of this book you will encounter proteins.

TESTS

RECALL TEST

1 Name two elements which are found in proteins but which are not found in carbohydrates or lipids.

_____ (1)

2 Draw a diagram to show the structural formula of an amino acid.

(1)

3 How many different types of amino acid are there?

_____ (1)

4 Which component of an amino acid is variable?

_____ (1)

5 What reaction is involved in the formation of a dipeptide?

_____ (1)

6 What bond is formed as a result of this reaction?

_____ (1)

7 Define 'primary protein structure'.

_____

_____ (1)

8 Name two types of coiling or folding that are used to produce the secondary structure of a protein.

_____

_____ (1)

9 Name three types of bond that help to maintain the tertiary structure of a protein.

_____

_____

_____ (1)

10 Name a protein that shows quaternary structure.

_____ (1)

(Total 10 marks)

CONCEPT TEST

1 A simplified diagram of a protein is shown below:

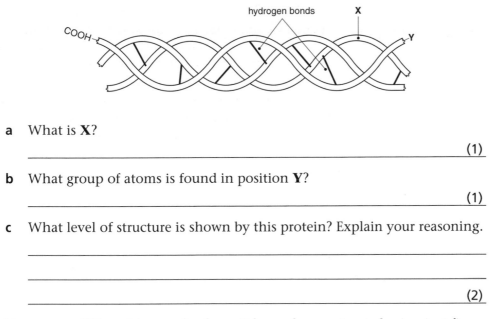

a What is **X**?

_____ (1)

b What group of atoms is found in position **Y**?

_____ (1)

c What level of structure is shown by this protein? Explain your reasoning.

_____

_____

_____ (2)

2 How many different types of polypeptide can be constructed using just five amino acids? Show your working in the space below.

(2)

3 **Cell surface receptors** are designed to recognize and bind on to specific messenger molecules, hormones, or neurotransmitters when they come into contact with a cell. Suggest why such receptors are likely to be made out of protein.

_____

_____

_____

_____ (4)

(Total 10 marks)

ENZYMES

● Right now, inside your body, there are millions of chemical reactions going on. In some of these reactions, small molecules are being stuck together to make bigger ones. Amino acids are being combined, for instance, to make new muscle protein. In other reactions, like the ones occurring in your gut, big molecules are being broken down.

● Most of these vital reactions would happen far too slowly, or not at all, if it weren't for **enzymes**. Enzymes are special molecules that act as **catalysts**. This means that they increase the **rate** of chemical reactions.

● Most reactions need an input of energy before they will get going. The energy that has to be put in to start the reaction off is called the **activation energy**. If an enzyme is present then less activation energy is needed to trigger the reaction, and so the reaction happens more easily. This is shown in the graph below.

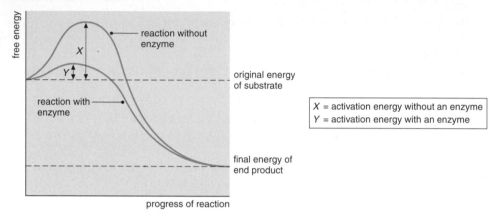

● *It is the special shape of an enzyme that allows it to reduce the activation energy needed to get a reaction going.*

● All enzymes are **globular proteins** and every enzyme has a small region somewhere on its globular surface known as an **active site**. This active site has a shape which is designed to fit exactly around one particular type of molecule. When an enzyme bumps into this kind of molecule, which is known as its **substrate**, the two join together to form an **enzyme–substrate complex**. While it is stuck in the enzyme's active site, the substrate is up to a million times more likely to undergo a reaction. Once the reaction has happened, the **end product** that has been produced is released out of the active site and the enzyme drifts off:

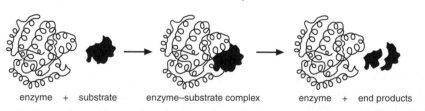

enzyme + substrate   enzyme–substrate complex   enzyme + end products

This explanation of how enzymes work is known as the **lock-and-key theory** of enzyme action, because the active site is like a lock into which only a specifically shaped key, or substrate, can fit.

● A more recent discovery is that, in many enzymes, the active site actually changes shape once the substrate has fitted into it. This process is called **induced fit**.

● A key point to note here is that enzymes are **specific**. Because the active site must fit the substrate perfectly, an enzyme can only catalyse one type of reaction. An enzyme with an active site designed to fit onto and break down sucrose cannot also fit onto and break down maltose. This means that there is a differently shaped enzyme for each one of the thousands of different types of reaction going on in the body.

- A variety of factors can affect the rate at which enzymes work. Graphs showing the effects of **temperature** and **pH** (acidity/alkalinity) on an enzyme-controlled reaction are shown below.

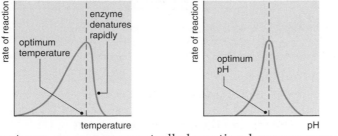

- At low temperatures, an enzyme-controlled reaction happens very slowly. You can see this on the graph. As the temperature is increased, the reaction speeds up. This happens because at higher temperatures all the molecules involved are moving around faster. This means that enzyme and substrate molecules will collide more often. Eventually the **optimum temperature** of the enzyme is reached, at which point the reaction is happening as fast as possible. At temperatures above the optimum, enzymes are shaken around so much that the hydrogen bonds which hold them in shape begin to break. When an enzyme's active site changes shape, so that it can no longer bind with its substrate, we say that it has been **denatured**. It is the irreversible denaturation of enzymes that causes the drop in reaction rate above the optimum temperature.

- As well as having an optimum temperature, enzymes also have an optimum pH at which they work best. This varies from enzyme to enzyme. Salivary amylase, found in the mouth, has an optimum pH of about 7. Pepsin, which has to work in the acidic environment of the stomach, has an optimum pH of about 2. If the pH diverges too much from the optimum, the presence of excess H^+ or OH^- ions can disrupt hydrogen bonding and cause denaturation in the same way that high temperature does.

- **Enzyme concentration** and **substrate concentration** are two other factors which can affect the rate of an enzyme-controlled reaction. Graphs are shown below.

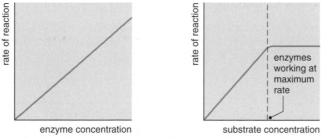

- The effect of enzyme concentration is relatively obvious. As more enzymes are added, the reaction goes faster. Substrate concentration is slightly more complicated. Initially, an increase in substrate concentration will increase the reaction rate, because there are more substrate molecules for the enzymes to work on. Eventually, however, the enzymes will all be working as fast as they can, and adding more substrate will have little or no effect. This is why the curve on the graph levels off.

- Finally, it is worth mentioning **inhibitors**. These are substances that reduce the activity of enzymes. Some are produced naturally in the body in order to control enzyme action. Some, like cyanide, are poisons that may be taken into the body by mistake.

- **Competitive inhibitors** have a similar shape to the substrate molecule. As a result they can block an enzyme's active site and prevent the real substrate from fitting in. **Non-competitive inhibitors** work by attaching onto the enzyme at some location other than the active site. This attachment changes the shape of the enzyme, including its active site, so that the substrate cannot fit.

TESTS

RECALL TEST

1 What are enzymes made of?

_____ (1)

2 Why are enzymes described as 'catalysts'?

_____ (1)

3 What is 'activation energy'?

_____ (1)

4 Arrange the following in the form of a flow chart:

enzyme + end product   enzyme + substrate   enzyme–substrate complex

(1)

5 Why are enzymes substrate-specific?

_____ (1)

6 What is 'induced fit'?

_____

_____ (1)

7 What bonds are broken when an enzyme is denatured?

_____ (1)

8 Name two factors that can cause enzyme denaturation.

_____ (1)

9 Name a substance that can act as an enzyme inhibitor.

_____ (1)

10 How do competitive inhibitors work?

_____

_____ (1)

(Total 10 marks)

CONCEPT TEST

1 A graph showing end-product accumulation in an enzyme-catalysed reaction at two different temperatures is shown below.

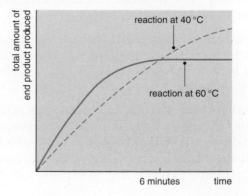

a Explain the difference in the shape of the two curves up to the 6 minute point.

_____

_____

_____

_____

_____ (2)

b Explain the difference in the shape of the two curves after the 6 minute point.

_____

_____

_____

_____ (3)

2 A graph showing the rate of an enzyme-catalysed reaction at different substrate concentrations, with and without inhibitors, is shown below.

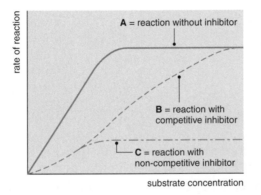

a Explain the shape of curve **A**.

_____

_____

_____

_____ (2)

b Explain the difference in shape of curve **B** and curve **C**.

_____

_____

_____

_____

_____

_____

_____

_____ (3)

(Total 10 marks)

CELLS

● A **cell** is a small bag, or sac, of water, surrounded by a thin membrane that separates it from the world outside. This membrane controls what substances are allowed to enter or to leave the cell. This means that vital molecules, such as nutrients and enzymes, can be trapped inside. Because the cell contains a high concentration of these molecules, chemical reactions can occur within it and life becomes possible.

● Some living organisms, such as the amoeba, consist of a single cell. Other organisms are **multicellular**. This means that they are built out of many cells. A human being contains billions of cells.

● The simplest type of cell is the **prokaryotic** cell. The first living organisms to evolve on this planet consisted of a single cell of this type. Modern-day **bacteria** are also single-celled prokaryotes. A simplified diagram of one is shown below.

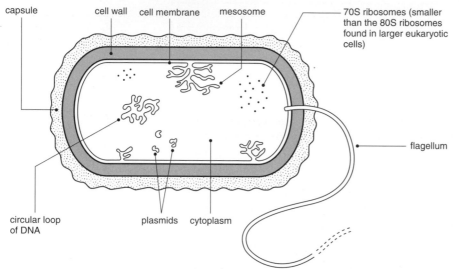

A bacterium

● The different structures that make up the cell are known as **organelles**. We have already mentioned the **cell membrane**, which you can see surrounding the bacterium. As well as controlling entry and exit, this membrane is folded in at key points, forming a **mesosome** to which vital enzymes can be attached. Outside the cell membrane there is a rigid **cell wall** and a **capsule**. These help to protect the bacterium. Inside the cell there is a big loop of **circular DNA**, and several smaller loops of DNA called **plasmids**. These control all the processes that go on inside the cell. The liquid inside the cell is known as **cytoplasm**, and floating around in this are **ribosomes**, which are involved in protein manufacture. Finally there is a tail, or **flagellum**, which the bacterium can beat from side to side in order to propel itself forward.

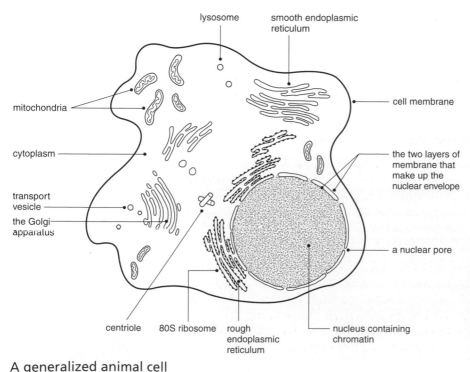

A generalized animal cell

● A prokaryotic cell is, in essence, a single space or compartment surrounded by a membrane. A more complex type of cell is the **eukaryotic** cell. As well as being much larger, eukaryotic cells contain membrane-bound spaces or subcompartments inside the main cell. Different types of chemical process occur in different subcompartments, or **membrane-bound organelles** as we call them. This localization of different chemical processes increases the efficiency of the cell.

● The most well-known types of eukaryotic cell are the **animal cell** and the **plant cell**.

● As was the case with the bacterial cell, both plant and animal cells consist of cytoplasm surrounded by a cell membrane. In addition to this, the plant cell has a rigid **cellulose cell wall** which provides it with support. Adjacent plant cells are connected by thin channels of cytoplasm which run through their cell walls. These are known as **plasmodesmata**.

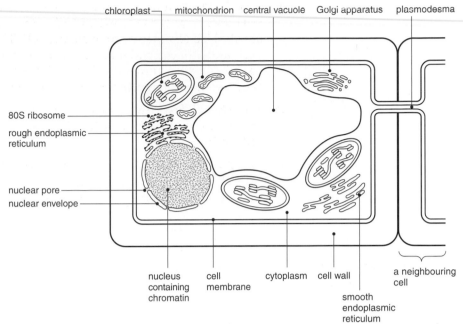

A generalized plant cell

- The DNA in a prokaryotic cell floats loose in the cytoplasm. In a eukaryotic cell, the DNA is contained inside a **nucleus**. The nucleus is a specialized subcompartment of the cell, surrounded by a **nuclear envelope**. This envelope consists of two membranes with fluid between them. It is penetrated, at numerous points, by **nuclear pores**, through which material can leave and enter the nucleus. The tangled mass of DNA inside the nucleus is known as **chromatin**. It is inside the nucleus that **transcription**, the first stage of protein synthesis, occurs (see unit 11).

- **Mitochondria** and **chloroplasts** are also specialized subcompartments. Like the nucleus, they are both surrounded by a double layer of membrane. Mitochondria are the site of respiration (see unit 8). Chloroplasts, which are absent from animal cells, are the site of photosynthesis (see unit 9).

- The **endoplasmic reticulum** is a system of membranes that divides the cytoplasm up into yet more compartments. In the cell diagrams, only a part of this system is shown. In reality, the entire cell is filled with folded sheets and passageways of membrane. The **rough endoplasmic reticulum** is described as such because of the thousands of tiny ribosomes attached to it. These ribosomes are described as **80S** and are slightly larger than the **70S** ribosomes that are found in prokaryotes. The rough endoplasmic reticulum is the site of **translation**, the final stage of protein synthesis (see unit 11). The **smooth endoplasmic reticulum**, which lacks ribosomes, is the site of lipid synthesis and reactions involving carbohydrates.

- Proteins produced in the rough endoplasmic reticulum are sent, eventually, to the **Golgi apparatus** (or **Golgi body**), a stack of flattened, membrane-bound sacs where proteins and many other types of molecule undergo their final modification before being packaged into **transport vesicles**. Transport vesicles are tiny sacs of membrane that are used to carry materials from one place to another inside the cell.

- Other membrane-bound compartments include **lysosomes**, which contain digestive enzymes, **peroxisomes**, which produce hydrogen peroxide, and the large **central vacuole** of a plant cell, which is filled with water, ions, and dissolved organic molecules. The only solid objects of any decent size that can be found in a eukaryotic cell are the **centrioles**. These protein tubes, which are only present in animal cells, play a vital role in cell division (see unit 13).

- As a final point, it is worth noting that the cell diagrams shown in this unit are only generalized cell diagrams, showing features common to the majority of cells. Animal and plant cells, in reality, come in a huge variety of different types, each type specialized to do a different job. Their shapes may vary, from the elongated shape of a nerve cell to the biconcave disc of a red blood cell, as may the type and number of organelles that they contain.

RECALL TEST

1 What is the main difference between eukaryotic and prokaryotic cells?

_____

_____ (1)

2 What is a mesosome?

_____

_____ (1)

3 What is a plasmid?

_____

_____ (1)

4 What are the channels of cytoplasm that run through a cellulose cell wall called?

_____ (1)

5 Name three organelles that are surrounded by a double membrane.

_____

_____ (1)

6 Where in a cell does transcription occur?

_____ (1)

7 What is the difference between rough and smooth endoplasmic reticulum?

_____

_____ (1)

8 What is the function of the Golgi apparatus?

_____

_____ (1)

9 What do lysosomes contain?

_____ (1)

10 What shape is a red blood cell?

_____ (1)

(Total 10 marks)

CONCEPT TEST

1 A metre (m) contains a thousand millimetres (mm).
A millimetre contains a thousand micrometres (µm).
A micrometre contains a thousand nanometres (nm).

The size of an animal cell and some of its contents, in micrometres and nanometres, is given below:

animal cell 20 µm; mitochondrion 1 µm; cell membrane (width) 8 nm; nucleus 5 µm; ribosome 15 nm.

Imagine an animal cell as big as a room (say 5 m across). How big would the various organelles inside it be? Show your working below.

Size of nucleus

Size of a mitochondrion

Size of a ribosome

Width of cell membrane

(4)

2 Complete the table below with a tick (✔) if the statement is true or a cross (✗) if it is not true.

	Contains chloroplasts	Has a cell wall	Contains ribosomes	Contains centrioles	Has a nucleus
Bacterial cell					
Plant cell					
Animal cell					

(3)

3 The mitochondria found in eukaryotic cells contain their own loop of DNA as well as 70S ribosomes. Given this information, suggest how mitochondria may have originated.

_____

_____

_____

_____

_____ (3)

(Total 10 marks)

THE CELL MEMBRANE

- All cells are surrounded by a **cell surface membrane**, or **plasma membrane** as it is sometimes called. This membrane separates the contents of the cell from the world outside and controls which materials are allowed to leave or enter the cell.

- Many of the organelles found in eukaryotic cells, such as the nucleus and the Golgi body, are surrounded by their own membranes (see unit 5). These have a structure similar to that of the cell membrane.

- The cell membrane consists of two layers of phospholipid molecules, a **phospholipid bilayer**, with various protein molecules embedded in it. A diagram is shown below.

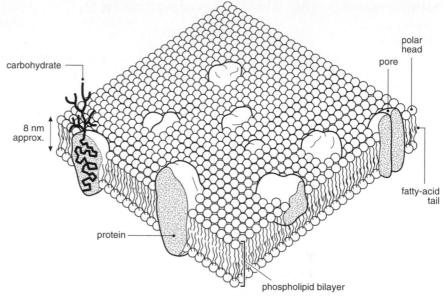

- You can see in the diagram that the heads of the phospholipid molecules are pointing outwards and that their tails are pointing inwards. There are important reasons for this arrangement.

- The heads of phospholipid molecules are charged (polar). This means that they can form hydrogen bonds with water (see unit 1). The consequence of this hydrogen bonding is that the heads of the phospholipids end up arranged on the surface of the bilayer, where they are in contact with water. We say that they are water-loving or **hydrophilic**. Because the heads are all crowded together near the surface of the bilayer, the fatty-acid tails, which are uncharged (non-polar), end up pointing inwards. We say that they are water-hating or **hydrophobic**.

- It is the hydrophilic nature of the heads and the hydrophobic nature of the tails that maintains the structure of the bilayer. There are no strong bonds tying all the phospholipids together. Instead, they can float around freely in the plane of the membrane, as long as their heads point out and their tails point in. Because of this, we describe the membrane as **fluid**.

- The situation is similar for the proteins embedded in the membrane. The parts of these proteins which are located close to water, among the phospholipid heads, are polar. The parts of the proteins located amongst the tails are non-polar. All the proteins can float around anywhere they want as long as they maintain this arrangement.

- Because the membrane is fluid and because, viewed from above, the proteins embedded in it look like the pieces of tile that make up a mosaic, we call this description of membrane structure the **fluid mosaic model**.

- The proteins embedded in the membrane do a variety of different jobs. Some are enzymes. Some are molecular pumps. Some act as receptors, detecting and binding on to molecules that bump into the surface of the cell. These proteins often have a bit of carbohydrate attached to them and are described as **glycoproteins.**

- One of the main ways that molecules can cross a membrane is by **diffusion**. Diffusion is the movement of molecules from a region where they are in high concentration to a region of lower concentration. If molecules are concentrated more on one side of a membrane than on the other, we say that there is a **concentration gradient** across the membrane. Molecules will continue to move down this concentration gradient until they are distributed evenly on either side of the membrane, at which point we say that **equilibrium** has been reached. Diffusion is a **passive** process, which means that it happens automatically, without the cell having to expend any energy.

- Uncharged molecules, such as fats or steroids, can diffuse across a membrane easily. They simply squeeze through in between the phospholipids. Charged molecules have more of a problem. Because of the charge they carry, they cannot enter the uncharged region of the membrane where all the phospholipid tails are. Instead, they must diffuse through special **channel proteins**. These proteins are like tunnels, lined with charge, which allow the passage of polar molecules.

- The rate of diffusion, for some molecules, is increased by the presence of **transport proteins** embedded in the membrane. These are proteins with a binding site, rather like the active site of an enzyme, which is designed to fit on to a specific type of molecule. Once they have combined with the appropriate molecule they flick it across to the opposite side of the membrane. This process is known as **facilitated diffusion**. Like simple diffusion, it is a passive process that requires no energy expenditure by the cell. It differs from simple diffusion because it is **selective**. Only molecules that fit into the binding site will be flicked across.

- Molecules of water move across a membrane by **osmosis**. This is really just a special case of diffusion, in which molecules of water move from where there is a *lower* concentration of solute molecules to where there is a *higher* concentration of solute molecules, across a partially permeable membrane.

- A big problem with diffusion and osmosis is that, in both processes, molecules can only go down a concentration gradient. Sometimes a cell may need to move molecules against a concentration gradient in order to accumulate them on one side of a membrane. This involves **active transport**.

- Active transport requires a **protein pump** which, like the transport proteins used in facilitated diffusion, has a specifically shaped binding site. Any molecule that fits into the binding site of the pump can be flicked across the membrane. The big difference between active transport and facilitated diffusion is that the protein pumps used in active transport have to work against the concentration gradient. Because of this, they require **energy** to drive them, which is why this type of transport is described as 'active'.

- A summary of all the different ways that molecules can cross a membrane is given below.

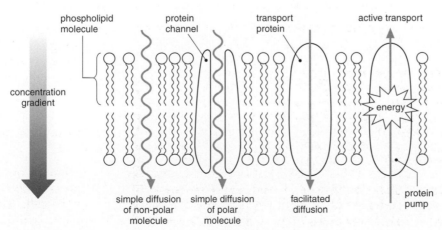

TESTS

RECALL TEST

1 What property of phospholipid molecules causes them to form up in a bilayer?

_____

_____ (1)

2 Why is a cell membrane described as 'fluid'?

_____

_____ (1)

3 List three functions performed by the proteins found in a cell membrane.

_____

_____ (1)

4 What is a glycoprotein?

_____ (1)

5 At what point will diffusion across a membrane cease?

_____ (1)

6 Why can't glucose molecules diffuse across a membrane by squeezing in between the phospholipids?

_____

_____ (1)

7 What type of molecule *can* cross a membrane in this way?

_____ (1)

8 Why are the transport proteins involved in facilitated diffusion described as 'selective'?

_____

_____ (1)

9 Why is osmosis described as a 'passive' process?

_____ (1)

10 Why is the active transport of molecules necessary?

_____

_____

_____ (1)

(Total 10 marks)

CONCEPT TEST

1 When viewed through an electron microscope, with sufficient magnification, the cell membrane appears as two dark bands separated by a light band, as shown in the diagram.

8 nm approx.

Suggest an explanation for this appearance.

_____

_____

_____ (2)

concentrations are in mol dm^{-3}

glucose 0.3	glucose 0.7
maltose 0.04	maltose 0.4
fructose 0.02	fructose 0.01

monosaccharide-permeable membrane

2 A diagram showing the concentration of various sugars on either side of a monosaccharide-permeable membrane is shown right.

  a In which direction will a net movement of water occur? Explain.

_____

_____ (2)

  b What other molecules will cross the membrane and in what direction will they move?

_____

_____ (2)

3 A graph showing how the rate of glucose uptake by a cell varies at different external glucose concentrations is shown below. Poisoning the cell has no immediate effect on the rate of glucose uptake.

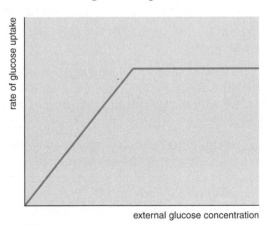

rate of glucose uptake

external glucose concentration

Use the information provided to deduce the type of cross-membrane transport involved. Explain your reasoning.

_____

_____

_____

_____

_____

_____ (4)

(Total 10 marks)

INVESTIGATING MOLECULES AND CELLS

- There are a variety of different techniques that scientists use to investigate molecules and cells. Many of these are simple enough to perform in a school laboratory.

- The presence of different types of molecule in a substance can be determined by using a series of straightforward chemical tests.

- **Benedict's test** is used to detect the presence of **reducing sugars**, which include all the monosaccharides and disaccharides (see unit 2) apart from sucrose.

- In Benedict's test, a couple of cm^3 of the solution to be tested is mixed together, in a test tube, with an equal volume of blue Benedict's solution. The test tube is then heated in a water bath. If reducing sugars are present a colour change should be observed. The degree of the colour change, from blue, through green, to brownish red, indicates roughly how much reducing sugar is present. The only problem with Benedict's test is that other reducing substances which are not sugars at all, such as vitamin C, may also cause a colour change.

- To test for the presence of sucrose, which is a **non-reducing sugar**, a modified version of Benedict's test is needed. The solution to be tested should first be boiled with dilute hydrochloric acid. This will cause any sucrose that it contains to be broken down into glucose and fructose. After the acid has been neutralized with sodium bicarbonate, a standard Benedict's test can be used to see if any glucose or fructose has been produced.

- To test for starch, the storage polysaccharide (see unit 2), a few drops of iodine should be added to the substance being tested. If the iodine changes from brown to blue-black then starch is present.

- The **Biuret test** is used to detect the presence of protein (see unit 3). A little potassium hydroxide is added to a test tube containing the solution to be tested. This should cause the solution to clear. A drop of copper sulphate is then dribbled in down the side of the test tube. If protein is present a dark blue ring should form on the surface of the solution and the solution should turn purple when shaken.

- A simple test for lipids (see unit 2) is to mix the solution to be tested with ethanol. After a few minutes of shaking, in a test tube, an equal volume of cold water should be added, and if lipids are present a cloudy white precipitate will be seen.

- A more thorough way of determining what types of molecule are contained in a particular substance is to use **chromatography**. This is a procedure that separates out different types of molecule according to how **soluble** they are in a particular **solvent**. If water-soluble molecules are being investigated then water is used as a solvent; for alcohol-soluble molecules, such as lipids, alcohol is used, etc.

- The substance being investigated is dissolved in the appropriate solvent and a drop of this solution is placed on a strip of filter paper. The paper is then suspended with one end submerged in solvent. As the solvent gradually soaks through the strip of filter paper it will meet and then pass by the point where the drop was placed, carrying with it any soluble molecules in the drop. Different types of molecule will be carried at different rates, according to their solubility, and will end up spread out across the filter paper.

- Each type of molecule is said to have a different R_f **value**, which is calculated as follows:

$$R_f \text{ value of molecule} = \frac{\text{distance moved by molecule}}{\text{distance moved by solvent}}$$

- A particular type of molecule, carried by a particular solvent, always has the same R_f value. This allows the molecules that have been spread out by chromatography to be identified.

- The contents of a cell are best investigated using **centrifugation**. This is a procedure in which different organelles are separated out in a rapidly spinning tube, according to their weight.

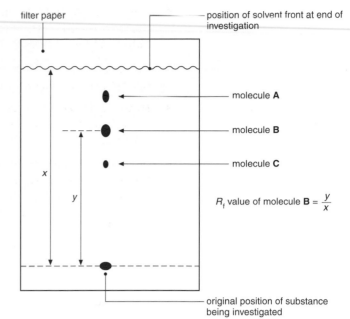

filter paper

- The organelles must first be released from within the cells that contain them. This is achieved by grinding up, or **homogenizing**, a particular type of tissue in a little bit of liquid. Plant material has to be ground with sand in order to break down its tough cellulose cell walls (see unit 5). The liquid in which the tissue is ground must be an **isotonic** solution, with exactly the same concentration as the liquid inside the organelles that are being released. If the grinding solution isn't isotonic, water may be sucked into, or out of, the organelles by osmosis (see unit 6), causing them damage.

position of solvent front at end of investigation

molecule **A**

molecule **B**

molecule **C**

R_f value of molecule **B** $= \dfrac{y}{x}$

original position of substance being investigated

A chromatography experiment

- After homogenization, the broken-up tissue is placed in a centrifuge tube and spun rapidly. The heaviest organelles, such as nuclei, will **sediment out** to form a pellet at the bottom of the tube. All the lighter organelles will be left floating around above the pellet in what we call the **supernatant**. If the supernatant is drawn off and then spun at an even higher speed, the next heaviest type of organelle, such as mitochondria, will sediment out. In this way, all the organelles that are found inside the cells of a given tissue can be separated out into different **fractions**.

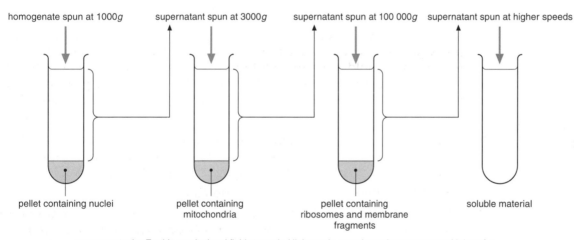

homogenate spun at 1000g supernatant spun at 3000g supernatant spun at 100 000g supernatant spun at higher speeds

pellet containing nuclei   pellet containing mitochondria   pellet containing ribosomes and membrane fragments   soluble material

g represents the Earth's gravitational field strength. Higher spin speeds produce greater multiples of g.

- To observe organelles or cells directly, a **microscope** is necessary. You have probably used a **light microscope** at school. An even more powerful magnifying tool is the **electron microsope**. This uses beams of electrons, rather than reflected light, to provide a detailed picture of very small objects.

- There are two main types of electron microscope. A **scanning electron microscope** fires an electron beam at an object and then collects any electrons that bounce off its surface. This provides a three-dimensional surface view of the object. A **transmission electron microscope** fires an electron beam at a thin slice of something which has previously been stained with heavy metals, and then collects any electrons which pass through it. This provides a two-dimensional view of the slice or **cross-section**.

- An electron microscope allows us to see much smaller things than with a light microscope. The electrons in an electron beam are much smaller than the photons in a beam of light. An electron microscope has a higher **resolution**. It can distinguish between smaller structures which are closer together. The disadvantage of electron microscopy is that objects have to be observed in a vacuum and thus cannot be viewed alive.

TESTS

RECALL TEST

1 What colour change is observed in a positive Benedict's test?

_____ (1)

2 What substance is used to test for the presence of starch?

_____ (1)

3 Name two chemicals that are used in the Biuret test.

_____ (1)

4 Write out the equation that is used to calculate the R_f value of a molecule in chromatography.

_____ (1)

5 Why do plant cells have to be ground with sand before centrifugation?

_____ (1)

6 What does 'isotonic' mean?

_____ (1)

7 What is the 'supernatant'?

_____

_____ (1)

8 Arrange the following centrifugation fractions in the correct order of separation, starting with the fraction produced at the lowest spin speed.

soluble material	nuclei	mitochondria	ribosomes
	+ mitochondria		+ membrane fragments

_____ (1)

9 What is the difference between resolution and magnification?

_____

_____ (1)

10 Why do electron microscopes have a higher resolution than light microscopes?

_____

_____ (1)

(Total 10 marks)

CONCEPT TEST

1 A variety of chemical tests were performed on two substances, **A** and **B**, and also on a mixture of **A** and **B** after it had been left for ten minutes. Substance **A** and the mixture of **A** and **B** were also investigated using a chromatography technique designed to separate out different types of carbohydrate. The results of the chemical tests and of the chromatography investigation are given in the following table.

	Biuret test	Benedict's test	Boiled with HCl + buffered + Benedict's test	Iodine test
A			✔	
B	✔			
A + B	✔	✔	✔	
✔ indicates a positive result				

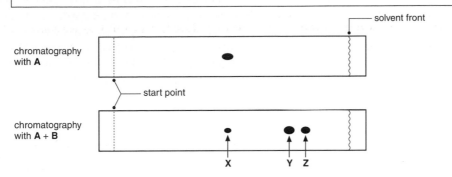

Using the information provided by the investigation, and explaining the reasons for your decision, identify:

a   Substance **A**;

_____

_____

_____ (2)

b   Substance **B**.

_____

_____

_____ (3)

c   The molecules found on the chromatograph in positions **X**, **Y**, and **Z**.

_____

_____

_____

_____ (2)

2   Three biological structures are listed below. In each case, suggest what type of microscope should be used to observe the structure.

a   The spines on the surface of a pollen grain;

_____ (1)

b   The beating heart of a water flea;

_____ (1)

c   The **lamella** (membranous fold) structure of a chloroplast.

_____ (1)

(Total 10 marks)

RESPIRATION

- All living organisms need a supply of energy. This energy is required to drive a variety of different processes, from the production of polysaccharides to the operation of protein pumps in the cell membrane.

- Living organisms gain the energy they need by breaking organic molecules down into simpler components. As the molecules are broken down the energy trapped inside them is released. This process is known as **cellular respiration**.

- The energy released by respiration is used to make molecules of **adenosine triphosphate**, or **ATP** for short. ATP contains three phosphate groups and it is produced by the **phosphorylation**, or addition of a phosphate group, to **adenosine diphosphate** (**ADP**), a molecule which contains two phosphate groups. The energy that is used to add the phosphate group to ADP ends up trapped inside the ATP molecule. The ATP molecule can then be transported to any part of the cell where energy is needed. When it arrives it will break down into ADP and a phosphate again, releasing the energy that is trapped inside it. In this way ATP molecules act as carriers of energy.

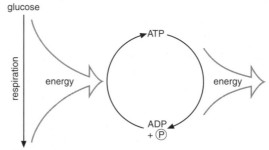

- The two main types of respiration are **aerobic respiration**, which requires the presence of oxygen, and **anaerobic respiration**, which does not. Anaerobic respiration is the most ancient and also the least efficient of the two processes. It can only be used to get energy out of carbohydrates, it generates less ATP than aerobic respiration, and it results in the production of poisonous by-products.

- The very first living organisms had to use anaerobic respiration because the early Earth didn't have an oxygen atmosphere. Today there are few entirely anaerobic organisms left, though there are many types of cell, such as yeast cells and muscle cells, that can operate anaerobically if there is no oxygen available.

- All of the biochemical steps involved in anaerobic respiration occur in the **cytoplasm** of the cell. The main process involved is known as **glycolysis**, a summary of which is shown left.

- Glycolysis begins with the phosphorylation of glucose, in which a phosphate from an ATP molecule is added to the original glucose molecule. The result of phosphorylation is that the ATP becomes an ADP, and the glucose becomes **glucose-6-phosphate** (the 6 refers to the fact that the phosphate is attached to the sixth carbon atom of the glucose molecule). Eventually, after another phosphate has been added, the original glucose molecule has acquired so much energy that it becomes unstable and splits into two molecules of **triose phosphate**. These triose phosphate molecules are highly reactive and very easy to get energy out of. It is worth putting energy in at the beginning of glycolysis, and actually using up ATP, in order to produce them.

- Once the triose phosphates have been produced, the real energy-generating part of glycolysis can begin. Each molecule is **oxidized**, by the removal of a hydrogen. The energy released by the oxidation of each triose phosphate molecule is enough to make an ATP.

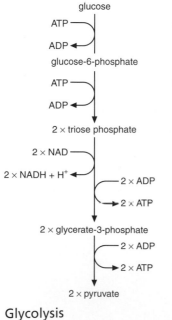

Glycolysis

- After oxidation, the triose phosphates become molecules of **glycerate-3-phosphate (GP)**. Each of these GPs is still carrying one of the phosphates that was added when the original glucose molecule was being split apart. These phosphates can now be reclaimed, to regenerate the two molecules of ATP that were used up in the first place.

- The net profit from the anaerobic breakdown of a glucose molecule, taking into account the ATP that has to be put in at the start, is 2 molecules of ATP.

- The two molecules of **pyruvate**, which are left at the end of glycolysis, still contain a lot of energy. Unfortunately, this energy cannot be accessed anaerobically. Instead, the pyruvates are used as a dumping ground. In the oxidation step of glycolysis, a molecule called **NAD** (nicotinamide–adenine dinucleotide) is used to carry hydrogen away, becoming **NADH + H⁺** in the process (NADH is the reduced form of NAD). NADH + H⁺ dumps its hydrogens, which have to be offloaded somewhere, onto pyruvate. It is this **reduction** of pyruvate that produces the poisonous by-products of anaerobic respiration: **lactic acid**, in animals, and alcohol, in plants.

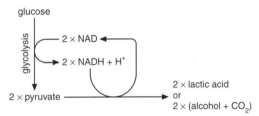

- Aerobic respiration, like anaerobic, begins with glycolysis. If oxygen is present, however, the pyruvate that's left at the end of glycolysis can be broken all the way down into CO_2 and H_2O, releasing its entire energy content in the process. For this to happen, the pyruvate has to enter a mitochondrion. The biochemical steps that occur inside a mitochondrion are shown below.

- The pyruvate that enters a mitochondrion is converted into **acetyl CoA** (coenzyme A), and this is fed into the **Krebs cycle**. This cycle involves a series of chemical reactions. In some of these reactions, carbon atoms are removed and released as CO_2. In one reaction ATP is produced directly. The main function of the cycle, however, is the **removal of hydrogens** from the molecules that are fed into it. These hydrogens are carried away by molecules of **NAD** and **FAD** (flavin–adenine dinucleotide).

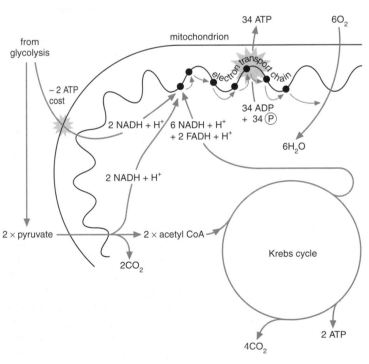

- All the hydrogens removed during the Krebs cycle, and those removed earlier, during glycolysis and the conversion of pyruvate to acetyl CoA, are then dropped off onto the **electron transport chain**. This consists of a series of molecules embedded in the inner membrane of the mitochondrion. As hydrogens are passed down this chain, from molecule to molecule, they release the energy that they contain. This energy can be used to make ATP from ADP, in a process known as **oxidative phosphorylation**. As they drop off the end of the chain, the hydrogens are picked up by oxygen, which, as a result, turns into H_2O.

- The complete aerobic breakdown of a glucose molecule gives a net profit of 36 molecules of ATP.

TESTS

RECALL TEST

1 How is energy transported from one place to another inside a cell?

_____ (1)

2 State three ways in which aerobic respiration is superior to anaerobic respiration.

_____

_____

_____ (1)

3 Where does glycolysis occur?

_____ (1)

4 What causes glucose to split apart into two molecules of triose phosphate in the early stages of glycolysis?

_____

_____ (1)

5 What is the biochemical advantage of lactic acid production in an anaerobically respiring animal cell?

_____

_____ (1)

6 What is the name of the molecule that is fed into the Krebs cycle?

_____ (1)

7 Where is the electron transport chain located?

_____ (1)

8 List three sources of the hydrogen that is fed into the electron transport chain in aerobic respiration.

_____

_____

_____ (1)

9 What is the terminal hydrogen acceptor of the electron transport chain?

_____ (1)

10 What is the net profit, in ATP, from the aerobic breakdown of a glucose molecule?

_____ (1)

(Total 10 marks)

CONCEPT TEST

1 The graph below shows how alcohol production, in an anaerobically respiring population of yeast cells, changes over time.

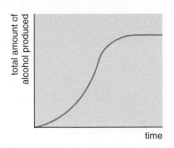

Explain the shape of the curve shown.

_____

_____

_____

_____

_____ (2)

2 The equations for aerobic and anaerobic respiration in ripening apples are shown below.

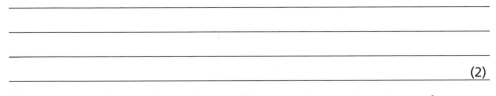

Aerobic $C_6H_{12}O_6 + 6O_2 \rightarrow 6CO_2 + 6H_2O$

Anaerobic $C_6H_{12}O_6 \rightarrow 2CO_2 + 2CH_3CH_2OH$

Despite what you might guess from looking at the equations, apples ripening in an oxygen-free environment actually produce more CO_2 per unit time than apples ripening in oxygen. Explain this difference in CO_2 output.

_____

_____

_____

_____

_____ (3)

3 Methylene blue is a blue solution that turns colourless when it picks up hydrogen. When mixed in a test tube with a population of respiring yeast cells, methylene blue will gradually lose its colour.

a Explain why this happens.

_____

_____

_____

_____ (3)

If the test tube is shaken vigorously, the methylene blue will go blue again.

b Explain.

_____

_____

_____ (2)

(Total 10 marks)

PHOTOSYNTHESIS

- **Photosynthesis** is a biochemical process in which molecules of **carbon dioxide** and **water** are combined to make **glucose**. The reactions involved are driven by **light energy**, and **oxygen** is given off as a by-product.

$$6CO_2 + 6H_2O \xrightarrow{\text{sunlight}} \underset{\text{glucose}}{C_6H_{12}O_6} + 6O_2$$

- Plants are renowned for their ability to photosynthesise, but there are many other organisms that can make glucose from CO_2 and water, including single-celled algae and cyanobacteria.

- All of these photosynthetic organisms are known as **autotrophs**, or self-feeders, because they use the glucose that they have produced by photosynthesis as the starting point for the manufacture of all the other types of organic molecule they need, including more complex carbohydrates, lipids, and proteins. **Heterotrophs** are organisms that feed on other organisms, gaining the organic molecules they need by stealing them, but without photosynthesis there would be no organic molecules in the first place.

- Photosynthesis is also the source of the Earth's oxygen atmosphere. The air of this planet would be unbreathable, by humans, if it weren't for all the oxygen that photosynthesis has released.

- A photosynthetic plant cell contains a number of membrane-bound organelles known as **chloroplasts** (see unit 5). It is inside these chloroplasts that the biochemical steps involved in photosynthesis occur.

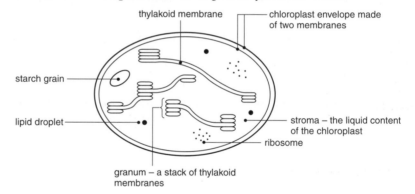

thylakoid membrane
chloroplast envelope made of two membranes
starch grain
lipid droplet
stroma – the liquid content of the chloroplast
ribosome
granum – a stack of thylakoid membranes

A chloroplast

- The aim of photosynthesis is to produce glucose, $C_6H_{12}O_6$. In order to do this the plant needs to obtain carbon, hydrogen, and oxygen.

- In the **light-independent** stage of photosynthesis, which occurs in the **stroma** of the chloroplast, the plant picks up, or **fixes**, CO_2. This provides the carbon and oxygen that is needed.

- In the **light-dependent** stage of photosynthesis, the plant splits molecules of H_2O apart to obtain the hydrogen that is needed. Molecules of **chlorophyll**, attached to the **thylakoid membranes** that make up the **grana** of the chloroplast, are used to absorb light energy. This energy is added to the hydrogens that have been split from water, so that they can be dropped off into the stroma to combine with the carbon and the oxygen. In this way, glucose can be produced.

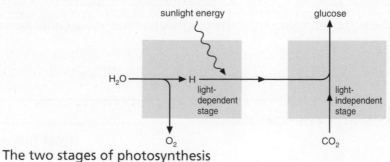

sunlight energy
glucose
H_2O
H
light-dependent stage
light-independent stage
O_2
CO_2

The two stages of photosynthesis

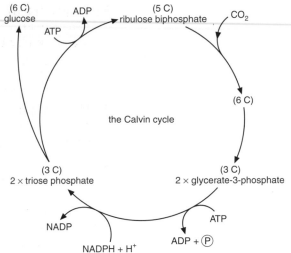

- The oxygen produced during the splitting of water, in the light-dependent stage, is not needed by the plant and is released as a by-product.

- The main biochemical process involved in the light-independent stage of photosynthesis is known as the **Calvin cycle**. The steps involved in this cycle are shown on the right.

- At the top right you can see CO_2 being picked up by a 5-carbon molecule called **ribulose biphosphate**. The 6-carbon molecule that is produced breaks down, almost instantly, into two 3-carbon molecules of glycerate-3-phosphate. It is at this point that hydrogen, which has been carried over from the light-dependent stage by a molecule called **NADP** (nicotinamide–adenine dinucleotide phosphate), is added in. Some of the molecules of triose phosphate that are produced as a result are used to make glucose, and some are used to regenerate ribulose biphosphate, which keeps the cycle going. At various points, energy, in the form of **ATP**, is required. This, like the hydrogen carried in by NADP, is also produced in the light-dependent stage.

- The steps involved in the light-dependent stage are shown right.

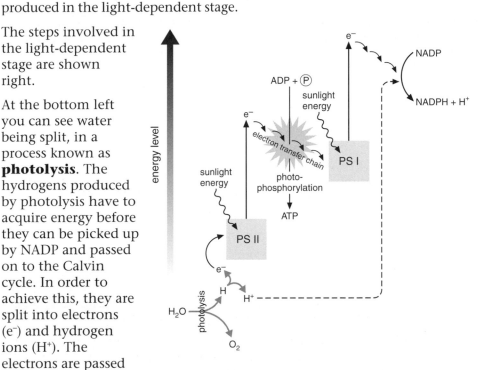

- At the bottom left you can see water being split, in a process known as **photolysis**. The hydrogens produced by photolysis have to acquire energy before they can be picked up by NADP and passed on to the Calvin cycle. In order to achieve this, they are split into electrons (e^-) and hydrogen ions (H^+). The electrons are passed to chlorophyll, where they acquire energy, and then recombined with the hydrogen ions afterwards, to produce high-energy hydrogens that can be picked up by NADP.

- The first collection of chlorophyll molecules that the electrons are passed to is known as **photosystem II**. Here, they are boosted to a higher energy level by the absorption of light. They still don't have enough energy, however, so they have to be passed down a chain of carrier molecules to yet another collection of chlorophyll molecules, **photosystem I**. Having received a final energy boost here, they can, at last, be recombined with the hydrogen ions that they were originally split from, and be picked up by NADP.

- You may remember that the Calvin cycle required ATP to drive a number of its steps. As electrons are passed down the chain of carrier molecules, from photosystem II to photosystem I, they release a little energy. It is this energy that is used to produce the ATP needed by the Calvin cycle, in a process known as **photophosphorylation**.

TESTS

RECALL TEST

1 Name three groups of organisms that can photosynthesise.

_____

_____ (1)

2 Why are photosynthetic organisms described as 'autotrophs'?

_____

_____ (1)

3 Where does the O_2 released by photosynthesis come from?

_____ (1)

4 Where, exactly, in a plant cell does the light-dependent stage of photosynthesis occur?

_____ (1)

5 What two products of the light-dependent stage are needed to keep the Calvin cycle running?

_____ (1)

6 What molecule is used in CO_2 fixation?

_____ (1)

7 How many carbon atoms does this molecule contain?

_____ (1)

8 What is photolysis?

_____ (1)

9 What is photophosphorylation?

_____

_____ (1)

10 Why are two photosystems necessary?

_____

_____

_____ (1)

(Total 10 marks)

CONCEPT TEST

1 In eukaryotic cells, chlorophyll and the other reactants involved in photosynthesis are contained within membrane-bound organelles called chloroplasts. Suggest one benefit of this arrangement.

_____

_____

_____

_____ (2)

2 DCPIP is a blue solution that turns colourless when it picks up high-energy electrons. If DCPIP is added to an illuminated test tube, containing isolated chloroplasts, it will lose its blue colour. It is, clearly, getting high-energy electrons from the chloroplasts.

a What is the original source of these electrons?

_____ (1)

b How do they become 'high-energy' electrons?

_____

_____

_____ (2)

c What would happen to these electrons normally, if DCPIP weren't present?

_____

_____

_____

_____ (2)

3 The graph below shows how the levels of glycerate-3-phosphate and triose phosphate inside a chloroplast change when it is moved into the dark.

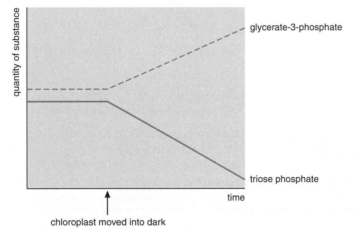

Explain the changes in concentration shown in the graph.

_____

_____

_____

_____

_____

_____

_____

_____ (3)

(Total 10 marks)

DNA

- All human beings grow from a single cell, or **zygote**, which is formed when a sperm and an egg fuse together. If we compare a human zygote with one that is going to grow into a camel or one that is going to grow into a chicken we will see no difference between them. They each have cytoplasm and mitochondria and all the other usual organelles. The only way these zygotes vary is that, within their nuclei, they contain slightly different versions of a molecule called **DNA**. The inheritance of human DNA will turn a cell into a human. Chicken DNA will produce a chicken.

- DNA stands for **deoxyribonucleic acid**. This tells us that DNA is a type of **nucleic acid** containing the sugar **deoxyribose**.

pentose sugar (deoxyribose)

phosphate group

nitrogen base
adenine } purine
or guanine } bases

or thymine } pyrimidine
or cytosine } bases

- All nucleic acids are built up out of **nucleotides** (see left). A nucleotide is a small molecule consisting of three components: a **phosphate group**, a **pentose sugar** (deoxyribose, if it's a DNA nucleotide), and a **nitrogenous base**.

- There are four different types of nitrogenous base that can form part of a DNA nucleotide: **adenine**, **guanine**, **thymine**, and **cytosine**. Adenine and guanine are large bases, made from two rings of carbon and nitrogen. They are known as **purines**. Cytosine and thymine are smaller bases, with one ring, known as **pyrimidines**.

- A condensation reaction can be used to join two nucleotides together. This results in a **phosphodiester bond** which links the phosphate of one nucleotide to the pentose sugar of the next. If many nucleotides are linked together in this way then a **polynucleotide** chain is produced (see below).

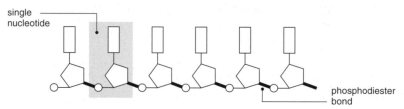

single nucleotide

phosphodiester bond

- A DNA molecule consists of two polynucleotide chains lying side by side. The chains are linked together by hydrogen bonds between their bases. Purine bases always link with pyrimidines. In fact the only possible pairings are adenine–thymine and guanine–cytosine. A pair of bases which can link together in this way are described as **complementary**. Diagrams of a DNA molecule are shown below.

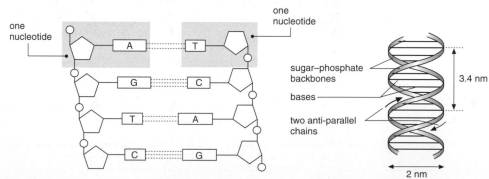

one nucleotide

one nucleotide

sugar–phosphate backbones

bases

two anti-parallel chains

3.4 nm

2 nm

- You can see from the main diagram that the two polynucleotide chains run in opposite directions. Because of this we describe them as being **anti-parallel**. What the main diagram doesn't show, because it's only two-dimensional, is that the chains are actually twisted around each other to form a spiral shape. A simplified view of this **double helix** is shown in the smaller diagram.

- In prokaryotic cells there is one long piece of DNA that forms a loop or circle. In the nucleus of a eukaryotic cell there are many long pieces of DNA, each of which is wound around a series of **histone proteins** to provide it with support. A piece of DNA, together with the proteins it's wound around, is known as a **chromosome**. Different organisms have chromosomes that contain different sequences of the four organic bases.

- Another important type of nucleic acid is **RNA** or **ribonucleic acid**. As its name implies, it contains **ribose** as opposed to deoxyribose sugar. Unlike DNA, RNA molecules consist of a single polynucleotide chain instead of two, and a pyrimidine base called **uracil** replaces thymine.

- As we mentioned at the start of this unit, it is the inheritance of human DNA that tells a cell to develop into a human. This inheritance is only possible because DNA is capable of **replication**; it can make copies of itself. These copies are what get passed from parent to offspring.

- During DNA replication the original double helix splits apart into two separate polynucleotide chains. A new complementary chain is then manufactured next to each of these original chains, so that each becomes a double helix again. The two pieces of DNA produced in this way are identical to the original. A simple diagram is shown to the right.

- Because the new pieces of DNA contain one old polynucleotide chain, from the original double helix, and one newly created polynucleotide chain, we call this method of replication **semiconservative**.

- The actual process is quite complex and involves a number of different enzymes. One type of enzyme is responsible for unzipping the double helix. Another type, called **DNA polymerase**, is designed to move along an unzipped polynucleotide chain, gradually assembling a new complementary chain beside it. The new chain is built out of individual nucleotides, which are found floating around in the nucleus. Every time the DNA polymerase passes an exposed base on the unzipped chain, it finds a free-floating nucleotide with a complementary base, and helps it to attach. As additional nucleotides arrive, the DNA polymerase joins them together with phosphodiester bonds. A summary is shown below.

original
polynucleotide
chain

new
complementary
polynucleotide
chain

Semiconservative
replication

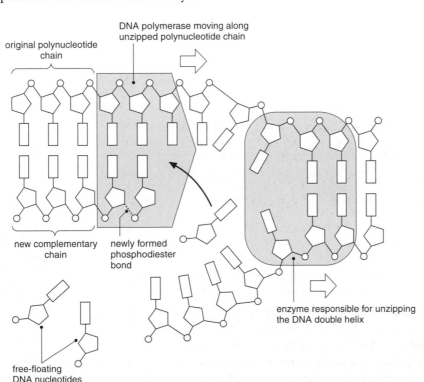

original polynucleotide
chain

DNA polymerase moving along
unzipped polynucleotide chain

new complementary
chain

newly formed
phosphodiester
bond

free-floating
DNA nucleotides

enzyme responsible for unzipping
the DNA double helix

- *It is the complementary nature of base pairing which guarantees that new pieces of DNA are identical to the original.*

TESTS

RECALL TEST

1 What does DNA stand for?

_____ (1)

2 What is a nucleotide composed of?

_____

_____ (1)

3 Name two purine bases.

_____ (1)

4 What bonds link nucleotides together to make a polynucleotide chain?

_____ (1)

5 What are the possible complementary base pairings that can be found in DNA?

_____ (1)

6 Why are the polynucleotide chains that make up a piece of DNA described as 'anti-parallel'?

_____

_____ (1)

7 What is a eukaryote chromosome composed of?

_____

_____ (1)

8 Give three differences between DNA and RNA.

_____

_____

_____ (1)

9 What enzyme is involved in DNA replication?

_____ (1)

10 What guarantees that DNA replication is accurate?

_____

_____ (1)

(Total 10 marks)

40

CONCEPT TEST

1 The proportions of the four organic bases in a piece of DNA from a squirrel, in a piece of DNA from a shark, and in a piece of human **mRNA** (messenger RNA) are given below.

	A	G	C	T/U
Squirrel DNA	29%	21%	22%	28%
Shark DNA	28%	21%	21%	30%
Human mRNA	40%	15%	30%	15%

a Explain why the proportion of A + G is approximately equal to the proportion of T + C in both pieces of DNA.

_____

_____

_____

_____

_____ (3)

b Explain why a similar equality does *not* exist between the proportion of A + G and the proportion of U + C in the RNA.

_____

_____

_____ (2)

c Despite being very different organisms, the squirrel and the shark have similar proportions of the four different bases in their DNA. Explain how this is possible.

_____

_____

_____

_____ (2)

2 In a famous experiment to determine how DNA replicates, two scientists called Meselson and Stahl grew cells in a medium containing ^{15}N, an isotope of hydrogen that is heavier than normal. The DNA in these cells became labelled with the ^{15}N and thus heavier than normal. When the cells were moved to a medium containing normal nitrogen (^{14}N), the DNA that they contained, after one generation, was found to be lighter than before, but still heavier than normal. Explain how this evidence helps to support the theory of semiconservative DNA replication.

_____

_____

_____

_____

_____

_____ (3)

(Total 10 marks)

PROTEIN SYNTHESIS

- In the last unit, we explained that it is the DNA inside a zygote that determines what type of organism it will develop into. DNA can exert this kind of influence because it controls **protein synthesis**, the process in which new proteins, including enzymes, are made inside a cell. By controlling enzyme production, DNA can determine what types of reaction happen in a cell and thus what it will develop into.

- Inside the nucleus of a eukaryotic cell there are many long pieces of DNA known as **chromosomes**. Each chromosome carries information about the design of thousands of different proteins.

- A **gene** is the name we give to a small length, or section, of chromosome that carries information about one particular polypeptide. If, as in the case of haemoglobin, a protein contains more than one type of polypeptide (see unit 3), then more than one gene will be involved in its production. Between genes, and making up most of the chromosome's length, there is a lot of **junk DNA** which doesn't do anything.

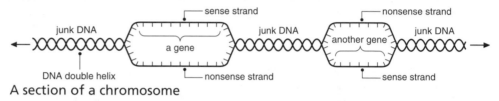

A section of a chromosome

- DNA is a double helix made up of two polynucleotide chains or strands. As you can see in the diagram, a gene consists of a small length of one of these two strands. The strand that forms the gene is called the **sense strand**. The complementary strand, which doesn't carry any useful information, is called the **nonsense strand**.

- Within the gene, it is the sequence of nitrogenous bases along the sense strand that contains information about how the polypeptide should be built. The order in which the bases are arranged determines the order in which amino acids will be combined together to make the polypeptide. This relationship between base sequence and amino-acid sequence is called **the genetic code**.

- There are twenty different types of amino acid which can be stuck together, in any order, to make a polypeptide. Unfortunately, there are only four different nitrogenous bases. This means that in order to **code** for all twenty amino acids the bases have to be read in groups of three. Each group of three bases is known as a **codon**.

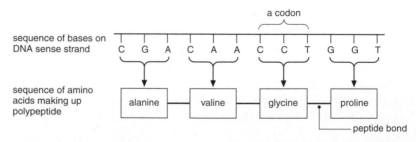

- There are sixty-four different codons altogether, which is more than enough to code for all twenty amino acids. Because of this, several different codons will often code for the same amino acid and, as a result, we call the DNA code **degenerate**. In addition to being a degenerate code, and a **triplet code**, because it's read in groups of three, the DNA code is also described as being **universal**. This is because the same codons are used, in all living organisms, to code for the same amino acids.

- As well as the codons which specify amino acids, there are also a few codons which are designed to mark where a gene begins and ends. These are known as **start codons** and **stop codons**. In a gene which codes for a polypeptide that is fifty amino acids long, we would expect a start codon, followed by fifty amino acid-specifying codons and then a stop codon, for a total of 156 bases. You can see from this that the length of a gene depends on the length of the polypeptide that it codes for.

● The actual process of protein manufacture in a cell is divided into two steps. In the first step, called **transcription**, the DNA double helix is partially unzipped and a temporary copy of the gene is made. This copy, in the form of **messenger RNA** (**mRNA**), is then passed out of the nucleus. In the second step, known as **translation**, the sequence of codons on the piece of mRNA is used as a guide for the assembly of amino acids into a polypeptide.

● The benefit of this two-step system is that many mRNA copies of a single gene can be produced and sent out of the nucleus. Each of these can then be used, repeatedly, to manufacture a large amount of the desired polypeptide.

● Transcription is under the control of an enzyme called **RNA polymerase**. The actions of this enzyme are summarized right.

● As you can see in the diagram, the RNA polymerase starts at the start codon and gradually moves along the DNA double helix towards the stop codon. As it goes along it unzips the double helix ahead of it. Free-floating RNA nucleotides are then attached, by complementary base pairing, to the exposed bases of the sense strand. In this way, a piece of mRNA which is complementary to the gene is gradually assembled. After it has passed by, the enzyme zips the double helix up again, and when it arrives at the stop codon the completed piece of mRNA floats off and out of the nucleus through a nuclear pore.

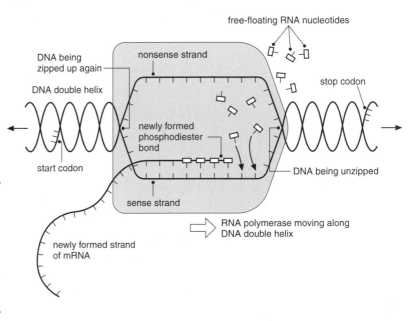

● Once it has left the nucleus, the piece of mRNA travels to the **rough endoplasmic reticulum** (see unit 5) where translation occurs. The process of translation is initiated when a ribosome, a large molecule composed of protein and **ribosomal RNA** (**rRNA**), attaches to one end of the mRNA strand. Having attached, the ribosome begins to move along the piece of mRNA, gradually assembling a polypeptide chain, as below.

● The amino acids which are needed to construct the polypeptide chain are carried into place by molecules of **transfer RNA** (**tRNA**). tRNA molecules are shaped like a clover leaf, with an amino-acid binding site at one end, and at the other end a group of three projecting bases known as an **anti-codon**. Every time the ribosome passes a codon on the mRNA, a tRNA molecule with a complementary anti-codon arrives and binds on to it. The amino acid carried by the tRNA molecule is then attached, by a peptide bond, to the growing polypeptide chain, and the tRNA molecule, having done its job, is allowed to float off again. The polypeptide chain grows in this way, amino acid by amino acid, until the ribosome finally reaches the end of the mRNA strand and the polypeptide is completed.

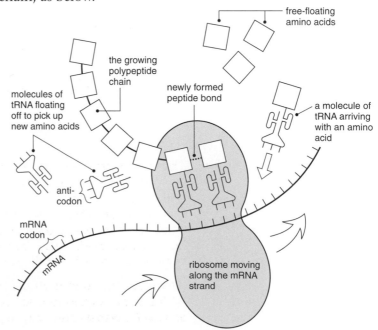

TESTS

RECALL TEST

1 Arrange the following in a simple flow chart.

translation DNA phenotype enzyme transcription chemical reaction

_____ (1)

2 What is a gene?

_____

_____ (1)

3 What is a codon?

_____

_____ (1)

4 How many different codons are there?

_____ (1)

5 Why is the genetic code described as 'universal'?

_____

_____ (1)

6 What enzyme controls transcription?

_____ (1)

7 Name two types of chemical bond that are formed during transcription.

_____ (1)

8 Where in a cell does translation occur?

_____ (1)

9 What is a ribosome made of?

_____ (1)

10 What is the role of tRNA molecules in translation?

_____

_____

_____ (1)

(Total 10 marks)

CONCEPT TEST

1 The base sequence of a particular gene and the polypeptide that the gene codes for are shown below.

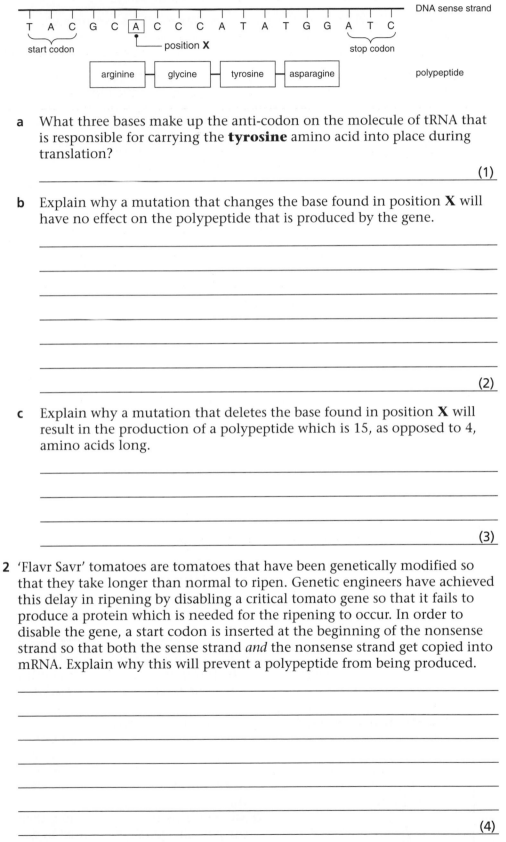

a What three bases make up the anti-codon on the molecule of tRNA that is responsible for carrying the **tyrosine** amino acid into place during translation?

_____ (1)

b Explain why a mutation that changes the base found in position **X** will have no effect on the polypeptide that is produced by the gene.

_____

_____

_____

_____

_____

_____ (2)

c Explain why a mutation that deletes the base found in position **X** will result in the production of a polypeptide which is 15, as opposed to 4, amino acids long.

_____

_____

_____ (3)

2 'Flavr Savr' tomatoes are tomatoes that have been genetically modified so that they take longer than normal to ripen. Genetic engineers have achieved this delay in ripening by disabling a critical tomato gene so that it fails to produce a protein which is needed for the ripening to occur. In order to disable the gene, a start codon is inserted at the beginning of the nonsense strand so that both the sense strand *and* the nonsense strand get copied into mRNA. Explain why this will prevent a polypeptide from being produced.

_____

_____

_____

_____

_____

_____ (4)

(Total 10 marks)

GENETIC ENGINEERING

- The DNA inherited by a living organism determines what type of creature it will become. If the DNA is altered, then a different type of creature will be produced. The artificial alteration of an organism's DNA is known as **genetic engineering** or **gene technology**. Scientists can use genetic engineering to produce new types of plant and animal.

- At the current level of technology, the only way to alter an organism's DNA is to add in new genes or lengths of DNA base sequence. DNA which has been altered in this way, with new bits added in, is known as **recombinant DNA**.

- The modification of crop plants to produce genetically modified food, or **GM food** as it is more commonly known, is big business at the moment. Genes have being added in to plants, such as rice, potatoes, and tomatoes, to produce pest and herbicide resistance, to delay ripening, and to improve flavour.

- Another current use of genetic engineering is in **bioproduction**. This is where a gene that codes for some useful substance is added in to the DNA of a microorganism. The microorganism will then act as a tiny factory, consistently churning out the desired substance. Microorganisms are used because they multiply so quickly. A single genetically modified bacterium will soon give rise to a whole vatful. Examples of bioproduction include bacteria engineered to produce **insulin**, which is used to treat diabetics, and yeast engineered to produce **chymosin**, which is used to clot milk in the dairy industry.

- A recent breakthrough is the treatment of human **hereditary diseases** using genetic engineering. Many hereditary diseases are caused by the inheritance of a single faulty gene. Examples include **cystic fibrosis**, **haemophilia**, and **sickle-cell anaemia**. If scientists can insert functioning genes into the affected cells of a patient then the symptoms of the disease may be removed. This is known as **gene therapy**.

- Before they can genetically modify an organism, genetic engineers must first obtain a copy of the gene that is going to be inserted into the organism. This is known as **isolating the gene**, because the gene has to be separated out from all the rest of the DNA in the cells that it is being obtained from. Once the gene has been isolated it can then be inserted into the target organism. Sometimes a **vector**, such as a plasmid or a virus, is used to carry the gene into the organism. Sometimes the gene is inserted directly. A variety of isolation and insertion techniques are shown below.

Isolation techniques	Insertion techniques	
	A The gene is inserted into a **viral** vector which then inserts it into the target cells. **OR**	
A DNA from an organism is chopped up using **restriction enzymes**. The fragment containing the gene is then separated out using **gel electrophoresis**. **OR**	**B** The gene is inserted into a plasmid vector which is then picked up by the target cells. **OR**	(can only be used to get genes into bacteria)
B Reverse transcriptase is added to the mRNA from an organism to produce **cDNA** (copy DNA) copies of the desired gene.	**C** The gene is inserted into the target cells using a fine syringe. This is known as **microinjection**. **OR**	(can only be used to get genes into large cells)
	D Metal fragments carrying the gene are fired at the target cells. This is known as **ballistic impregnation**.	(can only be used to get genes into cultures of single cells)

- **Restriction endonuclease enzymes** are enzymes that cut DNA. They are produced naturally by bacteria, which use them as a defence against viral infection. If a virus tries to take over a bacterium, the bacterium will use its restriction enzymes to chop up the invading viral DNA.

- There are many different types of restriction enzyme. Each type has a specially shaped active site (see unit 4) which recognizes and fits onto a particular sequence of DNA bases before making a cut in the DNA strand. Some examples are shown below.

- Genetic engineers can use restriction enzymes, which they have obtained from bacteria, to help them isolate a gene. First of all the DNA containing the gene is removed from the cells of the source organism and then restriction enzymes are added to it. The restriction enzymes chop the DNA up into pieces of different length, called **restriction fragments**. Some of these restriction fragments will contain the gene.

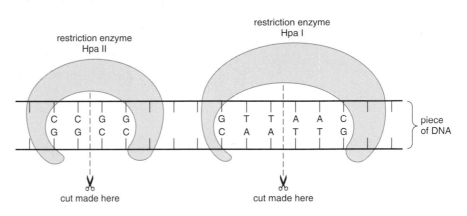

- To separate the restriction fragments containing the gene from all the other fragments, a process called **gel electrophoresis** is used. The restriction fragments are placed on one end of a piece of gel and an electric current is run across it. The electric current pulls the fragments, which are charged, through the gel, with the smallest and lightest fragments moving fastest and furthest. In this way the fragments can be separated out by length and the gene-containing fragments isolated.

- Another enzyme that can be used to isolate a gene is **reverse transcriptase**. This enzyme can be obtained from retroviruses, in which it naturally occurs. It reverses the normal process of transcription (see unit 11) by making a DNA copy of RNA. If pieces of mRNA are removed from the cells of an organism and reverse transcriptase is added to them, then pieces of **cDNA** (copy DNA) are produced. These pieces of cDNA will be identical to the gene that the mRNA pieces were originally transcribed from. In this way many copies of a gene can be obtained without any need to cut the DNA at all.

- There are many different techniques that can be used to insert genes into a host organism. **Viruses** have evolved, naturally, to insert their DNA into host cells (this is how they cause disease). This means that scientists can use viruses to carry useful genes into the cells of a target organism, though the virus's own DNA has to be disabled. **Plasmids** are loops of DNA that bacteria naturally swap between each other as a way of increasing genetic variation. If scientists can insert a useful gene into a plasmid, there is a chance that bacteria will take up the plasmid and the gene with it. To insert the gene, the plasmid is cut open using a restriction enzyme and then the gene is added. An enzyme, called **DNA ligase,** helps the cut ends of the plasmid to rejoin and form a loop again, but with the gene included.

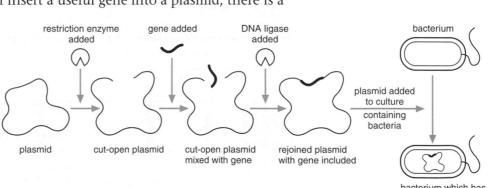

TESTS

RECALL TEST

1 What is meant by 'recombinant DNA'?

_____ (1)

2 Give an example of bioproduction.

_____

_____ (1)

3 What is meant by 'gene therapy'?

_____

_____ (1)

4 Why do bacteria produce restriction enzymes?

_____

_____ (1)

5 Why do different restriction enzymes cut DNA at different points?

_____

_____ (1)

6 What is 'cDNA'?

_____

_____ (1)

7 What is a vector?

_____

_____ (1)

8 Why will bacteria take up a gene more readily when the gene is part of a plasmid?

_____

_____ (1)

9 What does the enzyme DNA ligase do?

_____ (1)

10 What technique might genetic engineers use to insert genes into a human egg cell?

_____ (1)

(Total 10 marks)

CONCEPT TEST

1 A restriction enzyme was added to a sample of DNA and the restriction fragments produced were then separated out using gel electrophoresis. The result is shown below.

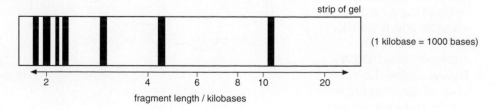

a Add an arrow to the diagram to show the direction in which the fragments moved during electrophoresis. (1)

b On the diagram you can see a series of vertical bands. What does each band consist of?

_____

_____ (2)

c If a different restriction enzyme had been used, a different banding pattern would have been produced. Why is this?

_____

_____

_____

_____ (2)

2 Some restriction endonuclease enzymes make a 'staggered' cut at the point on a piece of DNA that they recognize. An example of this is shown below.

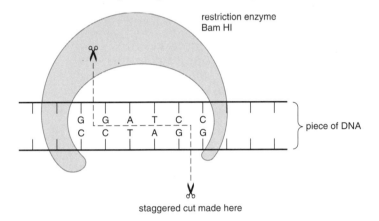

Pieces of DNA which have been separated by such a 'staggered' cut are described as having 'sticky ends'.

a Why is this?

_____

_____

_____

_____ (2)

b Why are restriction enzymes that leave 'sticky ends' useful to genetic engineers?

_____

_____

_____

_____

_____ (3)

(Total 10 marks)

MITOSIS

- All living organisms are built out of cells. When an organism grows or reproduces, new cells must be manufactured. These new cells aren't built from scratch, they are produced as the result of old cells splitting in two, in a process known as **cell division**.

- **Mitosis** is a type of cell division in which the two daughter cells produced are **genetically identical** to the original. In order to guarantee this, the DNA in the original cell must replicate, or double, before the cell divides.

- Mitosis is used to produce new cells during the **growth** of a multicellular organism and also during **repair**, when worn-out cells are replaced. Even if you have stopped growing, there are still mitotic divisions going on inside your body. When a single-celled organism such as an amoeba undergoes mitosis, to produce a pair of new and independent daughter cells, we say that **asexual reproduction** has occurred.

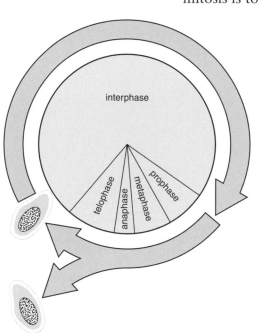

Mitosis

The cell cycle

- The best way to gain an overview of the stages, or **phases**, involved in mitosis is to take a look at something called the **cell cycle**, shown below.

You can imagine a cell travelling around this cycle, in a clockwise direction, passing through each of the named phases, one after another. When it reaches the end of telophase it splits in two and each of the daughter cells produced sets off around the cycle once again.

- The longest phase in mitosis is **interphase**. This is the phase between divisions, when the cell is doing whatever job it is designed to do. If you want to know what a cell looks like in interphase, just take a look at a standard cell diagram (see unit 5). The DNA appears as a tangled mass of chromatin, coiled up inside the nucleus.

- Because it is the phase that occurs between divisions, interphase is sometimes referred to as the **resting phase**. This term is a bit misleading. There are, in fact, a number of important events that have to occur in interphase in preparation for the next division. These include the growth of the cell, the storage of energy, and the replication of cell organelles. Most importantly of all, it is during interphase that the DNA replicates.

- The nucleus of a eukaryotic cell contains a number of pieces of DNA. Each of these long, double-helical pieces is known as a **chromosome**. During interphase, each chromosome undergoes semiconservative replication (see unit 10), to produce two copies of itself. These copies, which are known as **chromatids**, are held together, loosely, by a structure called a **centromere**. For historical reasons a pair of linked chromatids is referred to, during mitosis, as a single chromosome. This is a very confusing bit of terminology which may seem illogical, since a pair of linked chromatids includes two copies of the original chromosome, but it's just a bit of labelling that we're stuck with.

sister chromatids: each one is identical to the original chromosome

a 'chromosome' before replication = one piece of DNA

centromere

a 'chromosome' after replication = two pieces of DNA

- During interphase the replicating chromosomes are all tangled up together inside the nucleus and individual chromosomes cannot be distinguished. Once replication has finished, however, each chromosome begins to coil up or **condense** to form a small, compact bundle of DNA. The consequence of this condensation process is that individual chromosomes become apparent for the first time, at which point we say that the cell has entered **prophase**.

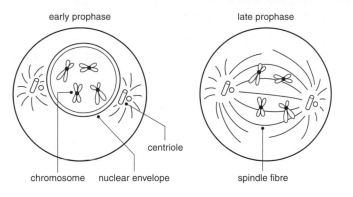

early prophase   late prophase

centriole

chromosome   nuclear envelope   spindle fibre

- The point of condensation is that, by untangling the chromosomes, it allows each one to be moved around independently of the others. During prophase a structure called the **spindle** is formed, which is used to move the condensed chromosomes about. The formation of the spindle is controlled by a pair of organelles known as **centrioles** (see unit 5). In the diagram of early prophase you can see these located at opposite ends, or **poles**, of the cell, where they are beginning to produce thin strands of protein known as **spindle fibres**. By late prophase the nuclear envelope has broken down and the spindle is complete, with spindle fibres spanning the cell from pole to pole.

- The chromosomes can now be moved. Spindle fibres latch onto them and drag them through the cytoplasm until they are lined up along the **equator** of the cell, with a spindle fibre attached to each centromere. At this point we say that the cell has entered **metaphase**. In the next phase, **anaphase**, the spindle fibres begin to shorten, splitting the chromosomes in two and pulling the separated chromatids towards opposite poles.

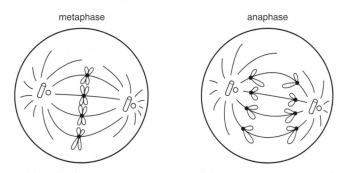

metaphase   anaphase

- On arrival at opposite poles, the two groups of chromatids uncoil again to become two separate blobs of chromatin. A nuclear envelope reforms around each group and the spindle, having done its job, breaks down. This finishing-off phase is known as **telophase**. The end result is two nuclei, each of which contains a set of chromosomes that is identical to the set that was in the original nucleus. The cell itself can now divide in two, by pinching in at the sides, in a process known as **cytokinesis**.

TESTS

RECALL TEST

1 Define 'mitosis'.

_____ (1)

2 List *three* functions of mitosis.

_____

_____ (1)

3 List the phases of mitosis, in the correct order, beginning with interphase.

_____

_____ (1)

4 In which phase do the chromosomes line up along the equator?

_____ (1)

5 In which phase does the spindle form?

_____ (1)

6 In which phase can sister chromatids be observed for the first time?

_____ (1)

7 What is a centriole?

_____

_____ (1)

8 What is a centromere?

_____

_____ (1)

9 After telophase it is no longer possible to distinguish individual chromosomes. Why is this?

_____

_____

_____ (1)

10 What is 'cytokinesis'?

_____

_____ (1)

(Total 10 marks)

CONCEPT TEST

Phase	Number of cells
interphase	16
prophase	12
metaphase	8
anaphase	3
telophase	7

1 The table shows the number of cells seen to be in each of the different phases of mitosis in a cross-section through the root tip of an onion.

Why are so few cells seen to be in anaphase?

_____ (1)

2 Many different events occur during interphase, including DNA replication. Towards the end of the phase there is actually twice as much DNA in the nucleus as there is at the start, and yet the scientists who first observed cells going through mitosis described the whole thing as just one phase: interphase. Why do you think this is?

_____

_____

_____ (2)

3 During mitosis we say that centrioles move to the 'poles' of the cell and that chromosomes line up along the 'equator'. Why do we use terms like 'poles' and 'equator' to describe positions in a cell?

_____

_____

_____ (2)

4 The graph below shows how the distance between various structures inside a cell changes during mitosis.

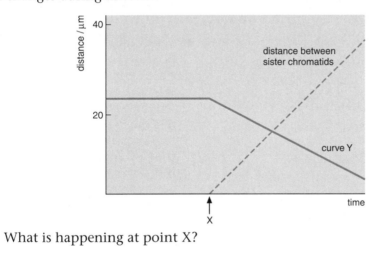

a What is happening at point X?

_____ (1)

b What does curve Y represent?

_____

_____ (2)

5 When scientists want to observe an organism's chromosomes they add a chemical called colchicine to a sample of cells from the organism that are about to go through mitosis. Colchicine inhibits the formation of spindle fibres. Why is it useful to add colchicine to mitotically dividing cells if you want to observe chromosomes?

_____

_____

_____

_____ (2)

(Total 10 marks)

MEIOSIS

- **Meiosis** is the division of a diploid parent cell to produce four, non-identical, haploid daughter cells.

- If you consider this definition you can see that there are three main ways in which meiosis differs from mitosis. Firstly, it results in the production of four daughter cells instead of two. Secondly, these daughter cells are genetically non-identical instead of being the same. Thirdly, and most importantly, the daughter cells produced by meiosis contain only half as much DNA as the parent cell. They are haploid, where the parent cell was diploid. Understanding what we mean by haploid and diploid is a key to understanding the whole of meiosis.

- A **haploid** cell contains a single complete set of chromosomes. Between them, these chromosomes carry all the genes needed by the cell, one gene for every polypeptide that has to be made. Some organisms, such as moss plants and male bees, are built entirely out of haploid cells.

- Most multicellular organisms, including human beings, are built out of diploid cells. A diploid cell contains two complete sets of chromosomes. This means that there are two genes for every polypeptide that needs to be made. There are two genes for eye colour, two genes that determine blood group, etc. A pair of chromosomes carrying genes for the same polypeptides, or characteristics, is known as a **homologous pair**.

A diploid set of chromosomes from the body cell of a fly. Several gene **loci** (the positions where they appear on a chromosome) have been labelled.

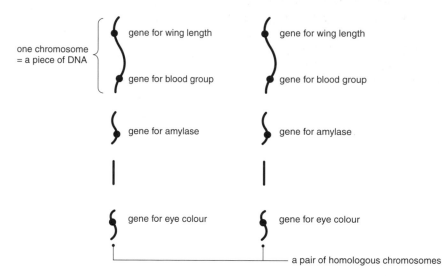

- You might be wondering, at this point, why diploid cells are necessary at all. Why have two sets of chromosomes when one will do? The answer has to do with sexual reproduction. When two organisms reproduce sexually, they each contribute genes to their offspring. The benefit of this gene mixing is that it generates variety. The best way to organize sexual reproduction is for each parent to be built of diploid cells containing two sets of chromosomes. When they want to reproduce, each of the parents creates a **gamete**, a haploid cell containing one set of chromosomes. When these gametes fuse, a new diploid organism is produced which contains chromosomes from both parents.

- You can see from the diagram left that the production of gametes requires a meiotic division, something that will produce a haploid cell from a diploid one. Gamete production is, in fact, the main function of meiosis. Because the gametes produced by meiosis contain only half the normal amount of DNA, meiosis is described as a **reduction division**.

- The process we call 'meiotic division' actually involves two cycles of cell division, one following the other. These are known as **meiosis I** and **meiosis II**. In each of these cycles you get the same sequence of phases as in mitosis: a prophase, a metaphase, an anaphase, and a telophase. A summary of the key events occurring in meiosis I and II is shown on the next page.

- Meiosis I starts, as does mitosis, with the interphase replication of DNA, in which each chromosome becomes a pair of chromatids. In prophase of meiosis I the DNA condenses, the nuclear envelope breaks down, the spindle fibres form, etc. The big difference between mitosis and meiosis I becomes apparent at metaphase. In mitosis the chromosomes form a single line down the equator, ready for chromatids to be pulled apart in anaphase. In meiosis I the chromosomes line up along the equator in homologous pairs and, in anaphase, it is these pairs that get pulled apart. No centromeres are split.

parents $2n$ $2n$
meiosis meiosis
gametes n n
fertilization
offspring $2n$

$2n$ = diploid
n = haploid

Sexual reproduction

- The daughter cells produced by meiosis I are officially haploid, since they only contain a single set of chromosomes. The problem is that these chromosomes still consist of pairs of chromatids. The aim of meiosis II is to separate these chromatids. As such, meiosis II is virtually identical to a mitotic division. The only real difference is that there is no replication of DNA at the start of meiosis II.

- Meiosis ends, as we explained at the beginning of the unit, with the production of four, non-identical, haploid daughter cells. These daughter cells are haploid because of the separation of homologous pairs in anaphase I and there are four of them because meiosis involves two complete cycles of cell division. Let us now consider why they are genetically non-identical.

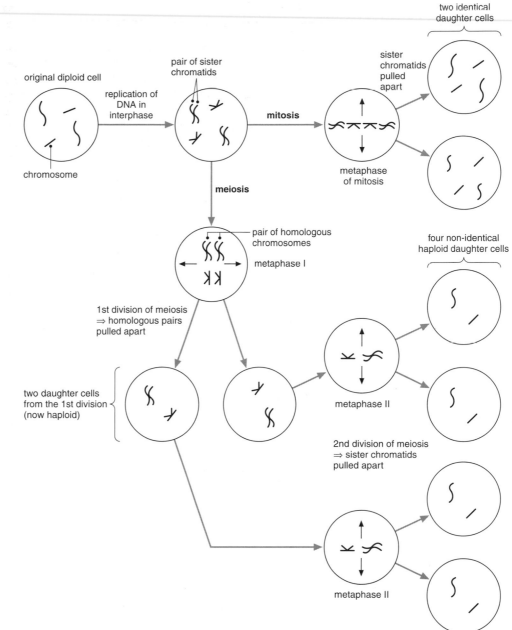

- Each of the daughter cells produced by meiosis is haploid. This means that it will receive one chromosome from every pair of homologous chromosomes in the diploid parent cell. Which chromosome it gets from each pair is, however, entirely a matter of chance, and depends on the way that the pairs line up in metaphase I. Because of this **independent assortment** of the parental chromosomes there are many different types of daughter cell that can be produced. Imagine a diploid parent cell with three pairs of chromosomes; A_1 and A_2, B_1 and B_2, and C_1 and C_2. A haploid daughter cell produced by the meiotic division of such a parent cell might end up with chromosomes A_1, B_1, and C_2 or with A_2, B_1, and C_1, or with any other combination.

- The variety of daughter cells that can be produced in meiosis is further increased by a process known as **crossing over**. During prophase and metaphase of the first meiotic division, when the homologous chromosomes are lying side by side they often become entangled. When they are pulled apart, in anaphase I, they end up exchanging lengths of DNA. This results in the production of chromosomes that contain new and unique combinations of genes.

TESTS

RECALL TEST

1 Define 'meiosis'.

_____

_____

_____ (1)

2 Why is meiosis described as a 'reduction division'?

_____

_____ (1)

3 What is the function of meiosis in human beings?

_____ (1)

4 Why do the chromosomes in a homologous pair differ from one another?

_____

_____ (1)

5 List *three* events that occur in prophase I of meiosis.

_____

_____ (1)

6 How do anaphase I and anaphase II of meiosis differ?

_____

_____ (1)

7 Why does meiosis result in the production of four daughter cells?

_____ (1)

8 A diploid parent cell contains two pairs of chromosomes: A and A' and B and B'. What combinations of chromosomes might be found in the daughter cells that result from the meiotic division of this parent cell?

_____ (1)

9 When does crossing over occur?

_____ (1)

10 How does crossing over contribute to genetic variation?

_____

_____ (1)

(Total 10 marks)

CONCEPT TEST

1 The graph below shows the changes in the quantity of DNA within a cell as it goes through a series of cell divisions.

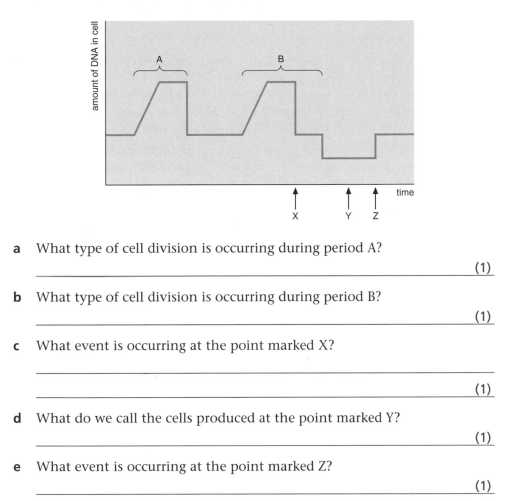

a What type of cell division is occurring during period A?

_____ (1)

b What type of cell division is occurring during period B?

_____ (1)

c What event is occurring at the point marked X?

_____

_____ (1)

d What do we call the cells produced at the point marked Y?

_____ (1)

e What event is occurring at the point marked Z?

_____ (1)

2 Complete the table below with a tick (✔) if the statement is true or a cross (✗) if it is not true.

	Centromeres are split	A spindle is formed	Homologous chromosomes pair up together	DNA replication occurs at the start of the division	Crossing over occurs
Mitosis					
Meiosis I					
Meiosis II					

(3)

3 If 6 chromosomes are observed in a cell at anaphase II of meiosis, work out how many chromosomes the cell would have contained at metaphase I.

(2)

(Total 10 marks)

57

AN INTRODUCTION TO MENDELIAN GENETICS

● We all inherit characteristics from our parents; eye colour, blood group, a particular shape of nose. Sometimes these characteristics seem to leap a generation. Sometimes two very tall parents can produce a short child. It gets quite complicated. **Mendelian genetics** deals with the laws that underlie the complex patterns of inheritance.

● The reason that children inherit their parents' characteristics is that they inherit parental genes. A gene is a small section of DNA that codes for a particular protein (see unit 11). The genes that an organism inherits from its mother and father together make up its **genotype**. It is the proteins produced by these genes that determine the organism's physical characteristics, its **phenotype**.

● Genes are found on long pieces of DNA known as chromosomes (see unit 11). The position at which a gene occurs on a chromosome is known as its **locus**. Humans inherit twenty-three chromosomes from each parent. Each set of twenty-three includes a locus for every gene needed to build a human being; a locus for a gene determining eye colour, a locus for a gene determining blood group, etc. Because we inherit a set of chromosomes from each of our parents, we end up with two genes for every characteristic: two genes for eye colour, two genes for blood group, etc. Other types of organism may have different numbers of chromosomes, some may inherit four from each parent, some may inherit fifty from each parent, but the same system applies.

● The gene for a particular characteristic that an organism inherits from its mother need not be identical to the one that it inherits from its father. It is possible, for example, to inherit a maternal eye-colour gene which codes for blue eyes and a paternal eye-colour gene which codes for brown eyes. Different versions of a gene, such as the blue and brown versions of the eye-colour gene, are known as **alleles**.

● How many different alleles, or versions of a gene, there are is something that varies from gene to gene. For most genes there is only one allele or version to be found, because the gene is vital for survival and any organism with a variant form will die. With other, less vital, genes there may be two or more different alleles. If there are more than two versions of a gene that can occur at a particular locus we say that the characteristic being coded for is controlled by **multiple alleles**. An example of such a characteristic is ABO blood group. There are three different ABO blood group alleles: I_A, I_B, and I_O. An individual can only inherit two of these, of course, one from their mother and one from their father, but that still results in a lot of possible genotypes: $I_A I_A$, $I_A I_O$, $I_B I_B$, $I_B I_O$, $I_A I_B$, and $I_O I_O$.

● Individuals who inherit two copies of the same allele, such as those with the blood-group genotype $I_A I_A$, are described as **homozygous**. Individuals who inherit a different allele from each parent, such as those with the genotype $I_A I_B$, are described as **heterozygous**. In the case of homozygous individuals it is obvious what phenotype is going to be produced, because both alleles code for the same thing. Someone with two brown eye-colour alleles will have brown eyes. Someone with the genotype $I_A I_A$ will have blood group A. With heterozygous individuals the determination of phenotype is more complicated and depends on the nature of the two different alleles that have been inherited. In some cases, one of the alleles will be **dominant** and the other **recessive**, in which case it is the dominant allele that will be expressed. The brown eye-colour allele, for example, is dominant to the allele for blue eye colour. If you inherit one of each, you end up with brown eyes. Another possibility is that the two alleles are **codominant**, in which case they will both have an influence on the phenotype of the heterozygous individual. The alleles I_A and I_B are codominant, which is why individuals with the genotype $I_A I_B$ are described as having blood group AB: neither allele dominates.

● Let us consider a mating, or **cross** as it is sometimes called, between a homozygous blue-eyed individual and a homozygous brown-eyed individual. In this case it is pretty easy to work out what kind of offspring will be produced. Every child will inherit a blue eye-colour allele from one parent and a brown eye-colour allele from the other. Since brown dominates blue, all the children will end up with brown eyes. This cross is illustrated below.

phenotype	blue eyes	×	brown eyes
genotype	bb		BB
gametes	ⓑ		Ⓑ

offspring: all Bb brown eyes

The symbol 'B' is used to represent the brown eye-colour allele and the symbol 'b' to represent the blue eye-colour allele, to remind us that brown is dominant to blue. The line labelled 'gametes' refers to the genotype of the eggs or sperm produced by the parents, which are going to fuse together to produce the offspring. Because these gametes are haploid (see unit 14) they only contain a single allele.

● Now let us consider a more complicated cross, between two brown-eyed individuals who are both heterozygous.

phenotype	brown eyes	×	brown eyes
genotype	Bb		Bb
gametes	Ⓑ or ⓑ		Ⓑ or ⓑ

offspring:

	Ⓑ	ⓑ	Punnett
Ⓑ	BB	Bb	square
ⓑ	Bb	bb	

3:1 ratio of brown eyes: blue eyes

This cross is complicated, as you can see, because each parent can produce two different genotypes of gamete. Which type they will contribute, in any given mating, is entirely random. There is a fifty per cent chance that they will contribute a 'B' gamete and a fifty per cent chance that they will contribute a 'b' gamete. The best way to predict what types of offspring are going to be produced, given this randomness, is to draw a **Punnett square**. When constructing a Punnett square, the gamete types that can be produced by one parent are written along the top and those that can be produced by the other parent are written down the side. The combinations of alleles that are produced by filling in the boxes show the possible genotypes of the offspring and the ratio in which they should occur. In the example above, the Punnett square tells us that out of every four children produced one should be homozygous brown eyed, two heterozygous brown eyed, and one homozygous blue eyed. This is only a statement of probability, or likelihood, of course. It is entirely possible that two heterozygous parents could produce four blue-eyed children in a row, just as it is possible to toss a coin eight times and get eight heads. It is just that these results are unlikely ones.

● Genetics problems can get a lot more complicated than the simple crosses that we have described here but, whether you are dealing with sex-linked genes or with **dihybrid** crosses, which involve two characteristics, the same basic rules apply.

TESTS

RECALL TEST

1 What does the word 'genotype' refer to?

_____ (1)

2 What does the word 'phenotype' refer to?

_____ (1)

3 What is a locus?

_____ (1)

4 What are 'alleles'?

_____

_____ (1)

5 Give an example of a human characteristic controlled by multiple alleles.

_____ (1)

6 What do we mean when we say that an organism is 'heterozygous' for a certain characteristic?

_____

_____ (1)

7 The allele for 'long tail' in mice is labelled s. The allele for 'short tail' is labelled S. Which allele is dominant and which recessive?

_____ (1)

8 In a certain species of plant, individuals who inherit two C^r alleles develop red flowers, while individuals who inherit two C^w alleles develop white flowers. Individuals that inherit a C^r and a C^w have flowers that are pink. What word would we use to describe the relationship between the C^r allele and the C^w allele?

_____ (1)

9 How many alleles for eye colour does a human sperm cell contain?

_____ (1)

10 The ratio of offspring types produced by two parents is often different from the ratio predicted by the laws of Mendelian genetics. Why is this?

_____

_____

_____ (1)

(Total 10 marks)

CONCEPT TEST

1 Coat colour in rabbits is controlled by a set of multiple alleles. These alleles are listed below.

    C produces normal coat colour
    c^{ch} produces 'chinchilla' coat colour when homozygous
    c^h produces 'himalayan' coat colour when homozygous
    c produces 'albino' coat colour when homozygous

Each allele in the sequence C, c^{ch}, c^h, c is dominant to the alleles that come after it in the sequence but recessive to the alleles that come before it. This is known as a **dominance hierarchy**.

a How many alleles for coat colour does each individual rabbit have?

                                                                  (1)

b What are the possible genotypes of a rabbit with a 'chinchilla' coat?

                                                                 (1)

c Work out the ratio of offspring phenotypes that you would expect to get from a cross between an albino rabbit and a heterozygous 'himalayan' rabbit. Indicate clearly, in your working, the genotypes of the parents, the gametes, and the offspring.

(3)

2 The diagram below shows a family tree in which the blood group of certain individuals has been labelled.

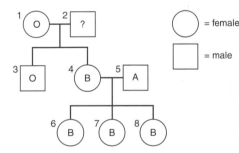

○ = female
□ = male

a What is the blood group of individual 2?

                                                                  (1)

b What is the genotype of individual 5?

                                                                    (1)

c What is the chance that the next child produced by individuals 4 and 5 will be another girl with blood group B? Show your working below.

(3)

(Total 10 marks)

EVOLUTION

- The first living organisms to appear on this planet were single-celled prokaryotes. No sooner had they appeared than they began to change and to diversify, adapting to the world around them. This process, continued over billions of years, has given rise to all the different types of living organism that we see today, from palm trees to tuna fish. We call this process **evolution**.

- The best way to get an overview of evolution is to take a look at a **phylogenetic tree**. This is a diagram that shows the history of, and relationships between, a number of different **species**.

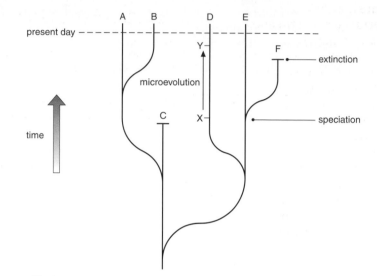

A phylogenetic tree

- A species is a group of organisms of one particular type that can breed together to produce fertile offspring but cannot breed with other organisms. Starlings, swifts, and swallows are all different species of bird because they cannot successfully interbreed. The history of six different species is shown in the phylogenetic tree, above. As you can see, all six are derived from a single common ancestor species. We could construct a phylogenetic tree for all living organisms, showing how they derive from the very first species of single-celled creature, but this would be a bit big.

- Important events shown in the phylogenetic tree include **speciation**, where a species splits in two, and **extinction**, where the last few members of a species die. What is less easy to show in the diagram is the gradual change in the nature of the organisms within each species, over time. This is known as **microevolution**. Imagine that, at point X on the diagram, the organisms of species D are all short legged and hairless. By the time we reach point Y, we might well find that they've evolved long legs and hairy coats.

- Microevolution is driven by a process known as **natural selection**. In every generation, within a species, new individuals are born and each of these new individuals has a slightly different genetic makeup. These genetic differences result from random changes in each individual's DNA, called **mutations**, and from the shuffling of genes that occurs during meiosis and sexual reproduction (see unit 14). Because they have different DNA, the new individuals will have different physical characteristics. Those individuals with characteristics which best suit them to the environment in which they live will have the best chance of surviving to reproduce. They will be 'naturally selected' to breed and pass on their genes. Individuals with genes that code for less useful characteristics will not survive to breed in such numbers. With this process of natural selection repeated every generation, useful genes will become more and more common and the individuals that make up the species will become better adapted to their environment. If the environment changes, new genes will be favoured by natural selection and the species will evolve to keep up with the environmental change.

- Microevolution can be illustrated using a graph, as right.

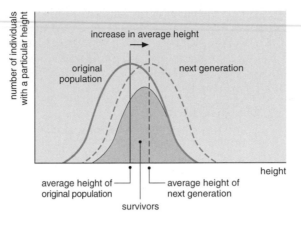

- The solid line on the graph shows the distribution of some characteristic, in this case height, in a particular population. As you can see, there are a lot of individuals of intermediate height and a smaller number of individuals which are shorter or taller than average. In this example we are going to presume that being taller gives individual organisms a better chance of surviving and breeding. The shaded area shows the height distribution of the individuals that manage to survive and, as you can see, they are mostly the tall ones. When these tall organisms breed together they produce a new generation which, because of mutation and gene shuffling, has a whole range of heights, as shown by the dotted line. You will notice, however, that the average height has increased because it is tall organisms that have been breeding and mixing their genes together. If being taller continues to be an advantage, the average height in this population will keep on increasing. The type of natural selection shown here is known as **directional selection** because it leads to a directional change in the average value of a characteristic.

- As well as microevolution, two other important evolutionary processes were illustrated in the phylogenetic tree: extinction and speciation. Extinction occurs when a species cannot evolve fast enough to keep up with a rapidly changing environment. Speciation is the process by which new species come into existence, when an old species splits in two to produce two groups of organisms that cannot interbreed. The most common type of speciation is known as **allopatric speciation**. It begins when part of a species becomes **geographically isolated** from the rest. There are all sorts of different ways in which this can happen. Some members of a bird species might colonize an island and then find themselves isolated from the mainland population. Changes in climate might cause a big forest to shrink into several smaller forests separated by grassland, isolating populations of a forest-dwelling species that were previously linked. Right at the start, the isolated populations are still considered to be part of the same species. This is because they are still capable of mating together, even if they do live in different places. Over time, however, it is likely that they will evolve to become more and more different from one another as they adapt to the different environments that they are isolated in. If they become so different from one another that they cannot interbreed, then speciation has occurred.

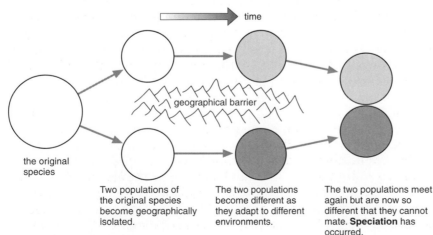

TESTS

RECALL TEST

1 What were the first living organisms to appear on this planet?

_____ (1)

2 Define 'species'.

_____

_____ (1)

3 What is shown by a phylogenetic tree?

_____

_____ (1)

4 What is meant by 'microevolution'?

_____

_____ (1)

5 Why do offspring have a different genetic makeup to their parents?

_____

_____ (1)

6 What do we mean when we say that fitter organisms are 'naturally selected'?

_____

_____ (1)

7 What does directional selection result in?

_____

_____ (1)

8 When does extinction occur?

_____

_____ (1)

9 What is 'speciation'?

_____

_____ (1)

10 What initial event is necessary if allopatric speciation is to occur?

_____ (1)

(Total 10 marks)

CONCEPT TEST

1 Though copper is poisonous to most plants, some species, such as the grass _Agrostis capillaris_, have evolved a degree of copper tolerance. The graphs below show the distribution of copper tolerance in two populations of _Agrostis_.

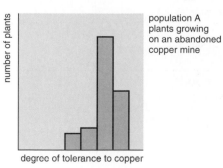

population A
plants growing
on an abandoned
copper mine

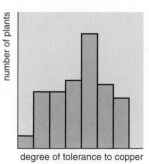

population B
seeds taken from
plants in population A
and then grown on
normal soil

a Explain the variation in copper tolerance found in population B.

_____

_____

_____

_____ (2)

b Explain why the plants in population A show a much narrower range of copper tolerance than those in population B.

_____

_____

_____

_____ (2)

c What evolutionary phenomenon is illustrated by the data?

_____ (1)

2 During the last ice age, ice sheets moved south to cover much of North America and then withdrew when the ice age came to an end. The map below shows the current distribution of three American bird species as well as a line marking the southernmost extent of the ice sheet.

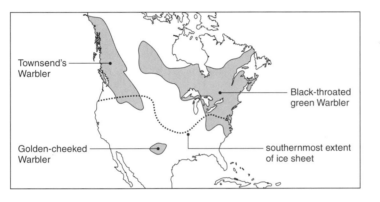

a Explain why the three species of Warbler are so similar in appearance.

_____

_____ (1)

b Explain why the three species differ in some of their physical characteristics.

_____

_____

_____ (2)

c Explain the role of the ice age in the development of the current diversity of Warbler species in North America.

_____

_____

_____

_____ (2)

(Total 10 marks)

THE DIVERSITY OF LIFE

● A **species** is a group of organisms that can breed with one another to produce fertile offspring. Lions and tigers are considered to be different species of big cat because, under natural conditions, they don't breed together. Labradors and Dalmatians, however, can breed together successfully, for which reason we consider them to be two different varieties of the same species: the domestic dog. During the history of this planet, the process of evolution (see unit 16) has given rise to a huge range of different species. Though many of these are now extinct, scientists estimate that there are least three million in existence at the moment, and new species are being discovered all the time.

● Every species of living organism known to science has an official **scientific name**. Scientists use these official names in the papers they write so that other scientists will know exactly what type of organism they are talking about. The bird that Americans call a 'robin' is quite different from the British 'robin', but this causes no problems for biologists, because the two species have different scientific names.

● The official system used to name species is known as the **binomial system**, because it involves giving every species a name that has two parts. The name given to the polar bear, for example, is *Ursus maritimus*. The first part of the name, *Ursus*, describes the **genus** to which the polar bear belongs. A genus is a group of closely related species. The genus *Ursus*, in this case, includes all the different species of bear. The second part of the name, *maritimus*, identifies exactly which species of bear is being referred to: *Ursus maritimus*, the polar bear, as opposed to *Ursus arctos*, the brown bear, or *Ursus americanus*, the black bear. The genus part of a scientific name always begins with a capital letter and the second part with a lower-case letter, and in printed material both parts are written in italics. An alternative to italics, if you are writing by hand, is to underline the name.

● As well as naming species, scientists also divide them up into different groups. This grouping of species has a number of advantages. A list of the groups gives a good overview of the diversity of life, and the fact that the species are grouped makes generalization easier. We can make a statement, such as 'mammals suckle their young', without having to list every single species of mammal. The science of dividing species up into groups, or **classifying** them, is known as **taxonomy**.

● The system of classification used by biologists is **hierarchical**. To start with, all known species are divided up into five **kingdoms**. Each of these five kingdoms is then divided up into a number of **phyla**, each phylum into a number of **classes**, each class into a number of **orders**, each order into a number of **families**, and each family into a number of genera. A genus, such as *Ursus*, which we mentioned earlier, may contain several different species. At each level in the hierarchy, it is the degree of relatedness between organisms that biologists pay attention to when they are deciding how to divide them up into groups. Humans, or *Homo sapiens*, to use our official species name, are put in the kingdom Animalia, for example, because we are more closely related to other animal species than we are to plants or fungi or bacteria. Within the animal kingdom, we are placed in the phylum Chordata because we are more closely related to other chordates such as birds and fish, which have backbones, than we are to beetles and worms and jellyfish. A full classification of *Homo sapiens* is shown below.

Kingdom	Phylum	Class	Order	Family	Genus	Species name
Animalia	Chordata	Mammalia	Primates	Hominidae	*Homo*	*Homo sapiens*
multicellular, heterotrophic organisms with cells that lack cell walls	animals with a dorsal nerve chord and, in most cases, a skeleton	warm-blooded and hairy chordates that suckle their young	monkeys and apes	bipedal, man-like apes	closely related tool-using hominids	

- In order to gain some idea of the diversity of life, let us look briefly at each of the five kingdoms and the types of organism that they contain.

- Kingdom **Prokaryotae**   This kingdom contains all the prokaryotic organisms, in other words the **bacteria** and their relatives the **cyanobacteria**. Bacteria are microscopic single-celled creatures with cells that lack the complex membrane-bound organelles found in eukaryotes (see unit 5). Despite the simplicity of their structure they are a highly successful group and are found in a wide range of habitats. Some, like *Lactobacillus*, are responsible for the decay of food. Some, like *Diplococcus*, which causes pneumonia, are disease-producing **parasites** that live inside, and feed off, other organisms. Some, like *Rhizobium* (see unit 27), are **mutualistic** and actually benefit the organisms whose bodies they inhabit. Cyanobacteria, which are similar in structure to bacteria, live by photosynthesis.

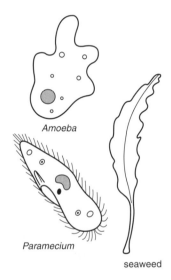

Prokaryotic organisms

- Kingdom **Protoctista**   This kingdom is a bit of a mixed bag, containing a range of eukaryotic organisms that wouldn't fit into the other kingdoms. These include single-celled **protozoans** such as *Amoeba* and *Paramecium*, which are mobile and heterotrophic (see unit 22), like animals, as well as seaweeds and other **algae**, which are photosynthetic like plants.

- Kingdom **Plantae**   This kingdom includes simple plants, such as **mosses** and **ferns**, as well as the more complex flowering plants or **angiosperms**. All of these are multicellular organisms with cellulose walls round their cells (see unit 5). With the exception of a few parasitic species, they gain the food they need by photosynthesis.

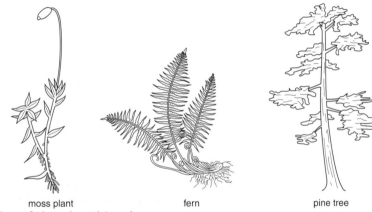

moss plant   fern   pine tree

Examples of the plant kingdom

Some protoctistans

- Kingdom **Animalia**   Animals are multicellular creatures whose cells lack cell walls. Though they are all heterotrophic, they vary hugely in complexity, from simple creatures such as jellyfish and worms to more complex types such as insects, fish, amphibians, etc.

- Kingdom **Fungi**   Though fungi are immobile, like plants, they are in fact heterotrophs. In most cases they feed on dead and decaying organic material. Some, like **yeast**, are single celled. Some, like **moulds** and **mushrooms**, are built out of long multinucleate strands called **hyphae**.

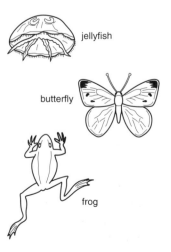

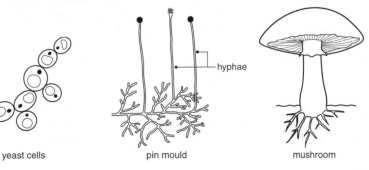

yeast cells   pin mould   mushroom

Examples of fungi

Examples of animals

67

TESTS

RECALL TEST

1 What is a species?

_____

_____

_____ (1)

2 Why is the official system used to name species known as 'the binomial system'?

_____

_____ (1)

3 What is the scientific name for the human species?

_____ (1)

4 What is 'taxonomy'?

_____

_____ (1)

5 In which kingdom are bacteria found?

_____ (1)

6 What is a parasite?

_____

_____ (1)

7 Give an example of a protozoan.

_____ (1)

8 What is an angiosperm?

_____ (1)

9 In which kingdom are algae found?

_____ (1)

10 Name a single-celled organism that is classified as a fungus.

_____ (1)

(Total 10 marks)

CONCEPT TEST

1 Palaeontologists are scientists who study fossils, the preserved remains of extinct organisms. What is the major problem that palaeontologists are faced with when they attempt to divide fossil organisms up into different species?

_____

_____

_____

_____

_____ (2)

2 Below is an incomplete classification of the wildcat, *Felis silvestris*. Fill in the gaps in the classification scheme.

Kingdom	Phylum		Order		Genus	Species Name
	Chordata	Mammalia	Carnivora	Felidae		*Felis silvestris*

(1)

3 DNA hybridization is a technique that can be used to reveal the similarities and differences between the DNA of two different organisms. Why might a taxonomist use this technique?

_____

_____

_____

_____ (2)

4 Complete the table below with a tick (✔) if the statement is true or a cross (✗) if it is not true.

	Contains single-celled organisms	Contains organisms built out of hyphae	Contains photosynthetic organisms	Contains eukaryotic organisms	Contains organisms with cellulose cell walls
Kingdom Prokaryotae					
Kingdom Protoctista					
Kingdom Plantae					
Kingdom Animalia					
Kingdom Fungi					

(5)

(Total 10 marks)

GAS EXCHANGE

● All living things have to exchange gases with their environment. Organisms that respire aerobically need to take up oxygen and get rid of carbon dioxide (see unit 8). Organisms that photosynthesise need to take up carbon dioxide and get rid of oxygen (see unit 9).

aerobic respiration	$6O_2$	+	$C_6H_{12}O_6$	$\rightarrow$	$6H_2O$	+	$6CO_2$
	oxygen in	+	glucose	$\rightarrow$	water	+	carbon dioxide out
photosynthesis	$6CO_2$	+	$6H_2O$	$\rightarrow$	$C_6H_{12}O_6$	+	$6O_2$
	carbon dioxide in	+	water	$\rightarrow$	glucose	+	oxygen out

● A single-celled creature, such as an amoeba, can exchange all the gases it needs by simple diffusion across its cell membrane (see unit 6). Since an amoeba uses up oxygen in respiration, there will always be more oxygen outside it than there is inside. Molecules of O_2 will enter the amoeba, automatically, as they diffuse down this concentration gradient. Carbon dioxide, which is produced by respiration inside the amoeba, will be at its highest concentration inside, so it will diffuse out.

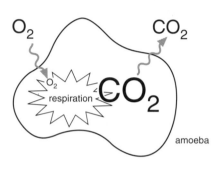

● For larger organisms, gas exchange presents more of a problem. A large organism cannot rely, like the amoeba, on the simple diffusion of gases across its normal body surface. The problem arises because as an organism gets larger it suffers from a decrease in its **surface area-to-volume ratio**. Consider the two cubes shown below.

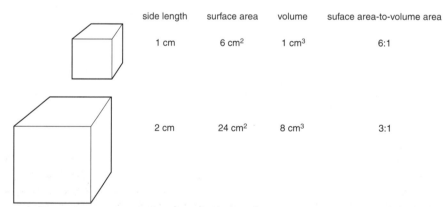

	side length	surface area	volume	suface area-to-volume area
	1 cm	6 cm^2	1 cm^3	6:1
	2 cm	24 cm^2	8 cm^3	3:1

The smaller cube has a surface area of 6 cm^2 and a volume of 1 cm^3. This means that its surface area-to-volume ratio is 6:1. The big cube has a larger absolute surface area, of 24 cm^2, but its surface area-to-volume ratio is only 3:1. This means that it has less surface area for each cm^3 of its volume. What is true for the cubes is also true for animals. As they get bigger, their increased volume means that they need to exchange more gas. The problem is that, with less and less surface area per unit volume, they simply cannot supply this demand by diffusion across their regular body surface.

● Large organisms overcome this problem by having a specialized **gas exchange surface**. This is a unique region of their body surface which is designed to maximize the diffusion of gases. Examples include the **lungs** in humans and the **gills** in a fish. Insects have a system of branching tubes called **tracheoles**. Plants have tiny pores, on the underside of their leaves, known as **stomata**.

- The rate of diffusion across a gas exchange surface can be determined by an equation known as **Fick's law**:

$$\text{rate of diffusion} = \frac{\text{surface area} \times \text{difference in concentration}}{\text{thickness of membrane}}$$

According to the equation, the rate of diffusion will be highest with a gas exchange surface that has a large surface area, is able to maintain a steep concentration gradient, and is thin. Bearing these factors in mind, let us take a look at the human gas exchange surface, the lungs.

- The lungs are, essentially, an infolding of the human body surface. From the back of the mouth a tube, known as the **trachea,** runs down into the chest cavity. The trachea divides into two smaller tubes called **bronchi** and these divide, repeatedly, to produce thousands of narrow branches known as **bronchioles**. At the end of every bronchiole is a tiny, thin-walled air sac or **alveolus**. Outside, and surrounding, each alveolus is a network of fine blood vessels, called **capillaries**.

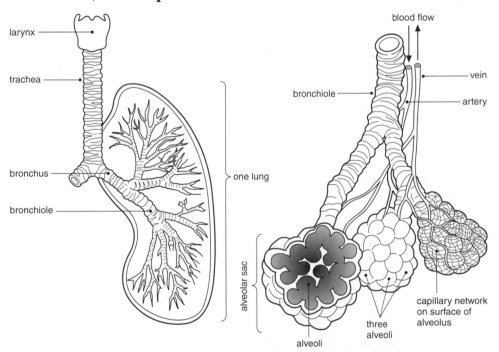

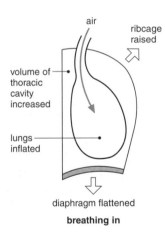

breathing in

- Gas exchange occurs across the walls of the alveoli. These are so thin that they allow molecules of O_2 to diffuse out of each alveolus and into the capillaries that surround it. The distance that an oxygen molecule has to move to pass through the wall of an alveolus and then the wall of a capillary is only $0.3\,\mu m$ (less than a thousandth of a millimetre). At the same time that oxygen molecules are diffusing out of an alveolus, molecules of CO_2 are diffusing in. The lungs provide a large surface area for this exchange of gases because, between them, they contain over 700 million alveoli.

- As oxygen molecules diffuse out of the alveoli, the concentration of oxygen inside the lungs begins to drop. In order to maintain a steep concentration gradient between the alveoli and the blood, stale air which has lost its oxygen must be regularly replaced with fresh air. This **ventilation** of the lungs is achieved by a muscular process known as **breathing**.

- The muscles that enable breathing do not act directly on the lungs. They are used, instead, to increase and decrease the volume of the **thoracic cavity**, the space inside the chest in which the lungs are suspended. If the **diaphragm** is flattened and the ribcage raised then the volume of the thoracic cavity increases. This lowers the pressure surrounding the lungs and causes them to inflate. If the diaphragm is allowed to bulge up again and the ribcage is lowered then the volume of the thoracic cavity will go down, causing a rise in thoracic pressure and causing the deflation of the lungs.

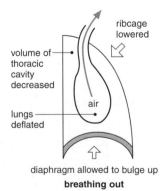

breathing out

How breathing ventilates the lungs.

TESTS

RECALL TEST

1 Name a biochemical process that uses up oxygen and releases CO_2.

_____ (1)

2 How does oxygen enter an amoeba?

_____

_____ (1)

3 What happens to an organism's surface area-to-volume ratio as the organism gets larger?

_____ (1)

4 What are 'tracheoles'?

_____

_____ (1)

5 What name is given to the tiny pores found on the underside of a leaf?

_____ (1)

6 State Fick's law.

_____ (1)

7 What are the 'bronchi'?

_____

_____ (1)

8 How do the lungs provide a large surface area for gas exchange?

_____

_____ (1)

9 Why is ventilation of the lungs necessary?

_____

_____

_____ (1)

10 What happens to the volume of your thoracic cavity when you breathe out?

_____ (1)

(Total 10 marks)

CONCEPT TEST

1 Unlike most worms, which exchange gases across their regular body surface, the lugworm, a particularly large variety of worm, has external gills spaced at regular intervals along its body.

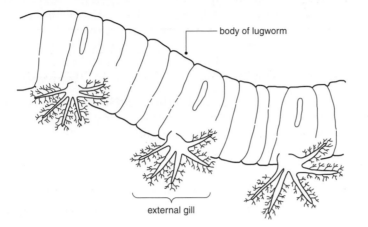

body of lugworm

external gill

Why does the lugworm require external gills?

_____

_____

_____

_____

_____ (2)

2 The gills of a fish consist of many thin flaps of tissue known as gill filaments. These blood-filled flaps project out from the inside of the fish's throat. The pumping action of the throat drives a stream of water into the fish's mouth, across the gills and then out of the gill slits in the side of its neck.

Explain, in terms of Fick's law, how a fish's gills help to maximize gas exchange.

_____

_____

_____

_____

_____

_____

_____

_____ (6)

3 Water, just like air, contains oxygen. If someone's lungs fill up with water, they can still absorb the oxygen out of the water and into their bloodstream. Given that this is the case, explain why humans cannot survive underwater.

_____

_____

_____

_____ (2)

(Total 10 marks)

TRANSPORT IN ANIMALS/THE HEART

● There are many different substances that have to be moved about inside the body of an animal, including gases, nutrients, and chemical messengers. Oxygen has to get from the gas exchange surface to the innermost parts of the animal. Food has to make its way from the place where it was digested to the places where it is going to be used or stored, etc.

● In a single-celled organism, such as an amoeba, the distances to be crossed are so small that substances can be left to move on their own by simple **diffusion**. In a large, multicellular organism this isn't good enough. If we relied on diffusion to get food from our gut to our innermost cells, we would soon be dead of starvation. Large organisms need a **transport system**.

● In animals, this role is performed by the **cardiovascular system**, a closed circuit of tubes filled with a circulating fluid. Any substances that are dropped off into this fluid are carried rapidly around the body. The fluid, as you may have guessed, is **blood** and the muscular pump which keeps it moving is known as the **heart**.

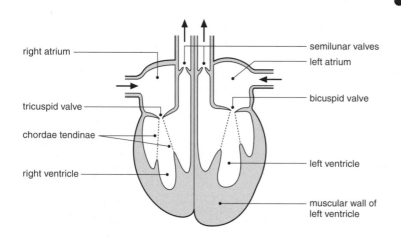

right atrium

tricuspid valve

chordae tendinae

right ventricle

semilunar valves

left atrium

bicuspid valve

left ventricle

muscular wall of left ventricle

● Human beings have what is known as a **double circulation. Oxygenated** blood, from the lungs, is passed to the left side of the heart, from where it is pumped to the body. **Deoxygenated** blood, returning from the body, is passed to the right side of the heart, from where it is pumped back to the lungs. Though the two sides of the heart lie next to each other and beat together, each one is really a separate pump, dealing with a separate flow of blood. A simplified diagram of the human heart is shown left.

The heart in the diagram is drawn as if it belonged to a person facing you. This means that the left side of their heart is on the right side of the page.

● Each side of the heart contains two chambers, an **atrium** and a **ventricle**. Blood enters the atrium first, where it accumulates for a while before being passed on to the ventricle, which is the main pumping chamber. The ventricle has thick muscular walls which can contract strongly, squirting blood out of the heart.

● When the ventricle contracts there is a danger that some blood may be squeezed in the wrong direction, back into the atrium. To prevent this from happening there is a one-way valve located between the atrium and the ventricle. On the left side of the heart this atrioventricular valve consists of two flaps of tissue and is known as the **bicuspid valve**. On the right side it consists of three flaps and is known as the **tricuspid valve**. The free edges of the flaps that make up an atrioventricular valve are anchored to the walls of the ventricle by thin fibrous cords called **heart strings** or **chordae tendinae**. These heart strings prevent the flaps from opening backwards so that although blood can pass from atrium to ventricle, it can't go back from the ventricle into the atrium. As well as the atrioventricular valve, there is also a smaller, stringless, valve that guards the exit from the ventricle. This **semilunar valve** is designed to prevent blood running back into the ventricle once it has been squeezed out.

● The sequence of muscular contractions and valve movements involved in the completion of a single heartbeat is known as the **cardiac cycle.** The main stages in this cycle, as seen in the right side of the heart, are shown in the diagram (top right).

● During **atrial systole** the atrium contracts, squeezing the blood that it has accumulated into the ventricle. This is followed by **ventricular systole**, in which the ventricle contracts. As the walls of the ventricle begin to close in, blood is forced back towards the atrium, closing the tricuspid valve. It is the impact of blood against the flaps of the tricuspid that causes the first heart sound or **lub**, as it is sometimes known. Eventually, as the ventricle continues to contract, the pressure inside the chamber reaches a high enough level to force open the semilunar valve, and blood is squirted out of the heart. When this blood tries to run back into the empty ventricle it encounters, and closes, the semilunar valve, causing the second heart sound or **dup**. Both atrium and ventricle now enter a period of relaxation, or **diastole**, during which they gradually refill with venous blood.

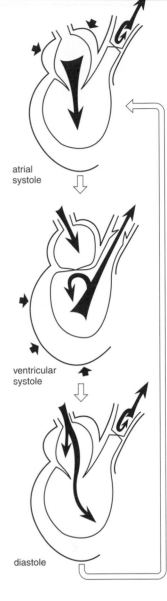

atrial systole

ventricular systole

diastole

- While the events that we have just described are happening on the right side of the heart, an identical sequence of events is happening on the left. The only significant difference between the two sides is that the left ventricle has a thicker muscular wall than the right ventricle and can generate more pressure when it contracts. The left ventricle needs to pump harder because it has to drive blood all the way round the body, while the right ventricle only needs to drive blood to the lungs.

- If the heart of a mammal is removed and placed in an oxygen-rich solution it will continue to beat for quite a while. It can continue to do this, without being stimulated by nerves, because it contains its own built-in mechanism for triggering muscle contraction. We say that it is **myogenic**. There are nerves running to the heart, but these are only needed to adjust the rate of the heartbeat.

- The initiation of a cardiac cycle is triggered by a small structure called the **pacemaker**, or **sinoatrial node**, which is located in the wall of the right atrium. This node generates a wave of electrical activity which spreads out through the muscular walls of the atria, causing them to contract from the top downwards, so

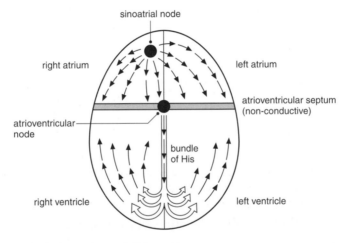

Controlling the cardiac cycle

that blood is squeezed into the ventricles. If this wave of electrical activity were allowed to keep on spreading it would next reach the upper walls of the ventricles, causing them to contract from the top downwards. This would result in blood bursting out of the bottom of the heart. To prevent such a disaster from happening there is a non-conductive layer, called the **atrioventricular septum**, separating the atrium from the ventricle. The only way that the wave of electrical activity can cross this layer is by passing through the **atrioventricular node**. As soon as it has passed through this node and into the ventricle it encounters the **bundle of His**, a structure composed of rapidly conducting **Purkinje fibres**. These carry the electrical wave to the bottom of the ventricles, from where it can spread upwards, causing an appropriate bottom-to-top ventricular contraction.

- As well as being myogenic and able to conduct electricity, heart muscle, or **cardiac muscle** as it is officially known, is also unique in its ability to resist fatigue. In a human lifetime the heart may beat as many as two billion times, and pump more than two million litres of blood, and through all this it never once grows tired.

TESTS

RECALL TEST

1 Which chamber pumps deoxygenated blood out of the heart?

_____ (1)

2 Where is the bicuspid valve located?

_____ (1)

3 What is the function of the chordae tendinae?

_____

_____ (1)

4 What causes the first heart sound or 'lub'?

_____

_____ (1)

5 What happens during diastole?

_____

_____ (1)

6 Why does the left ventricle need to generate more pressure than the right?

_____

_____

_____ (1)

7 Where is the sinoatrial node located?

_____ (1)

8 What is the function of the atrioventricular node?

_____

_____ (1)

9 What is the bundle of His made of?

_____ (1)

10 List three unusual features of cardiac muscle.

_____

_____ (1)

(Total 10 marks)

CONCEPT TEST

The graph shows the changes in blood pressure that occur in the left side of the heart during a single cardiac cycle. Blood leaving the left ventricle passes through the aorta.

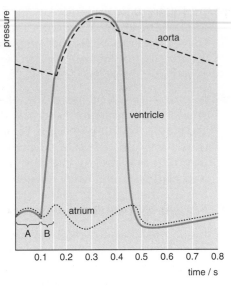

1 During the period labelled A on the graph, the atrioventricular valve is open.

   a Explain how someone could have worked this out using the information in the graph.

     _____

     _____

     _____

     _____ (2)

   b What is the official name given to the period labelled A on the graph?

     _____ (1)

2 During the period labelled B on the graph, there is a rise in ventricular pressure.

   a What is the cause of this pressure rise?

     _____ (1)

   b Is blood entering or leaving the ventricle during period B? Explain your reasoning.

     _____

     _____

     _____

     _____ (2)

3 Use the information in the graph to work out when the second heart sound, or 'dup', occurs. Explain your reasoning.

   _____

   _____

   _____

   _____

   _____ (3)

4 How would a graph showing pressure changes in the *right* side of the heart differ from the one shown above?

   _____ (1)

                        (Total 10 marks)

TRANSPORT IN ANIMALS/THE BLOOD

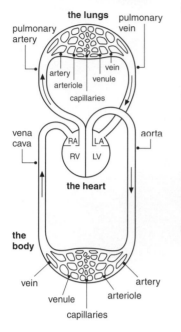

- In the last unit we looked at the heart, the muscular pump that drives blood around the body. An equally important part of the cardiovascular system is the set of tubes, or **blood vessels**, through which the blood travels. There are five main types of blood vessel: **arteries**, **arterioles**, **capillaries**, **venules**, and **veins**.

- Arteries carry blood away from the heart. This means that they have to receive high-pressure blood that has just been pumped out of the ventricles. To handle this high-pressure blood without bursting, they have thick muscular walls. The artery leaving the left side of the heart is known as the **aorta**. It branches, eventually, into many smaller blood vessels known as **arterioles**, which carry oxygenated blood to all parts of the body. The artery leaving the right side of the heart is known as the **pulmonary artery**. It branches into arterioles which carry deoxygenated blood to the lungs.

- Having reached the tissues, or the lungs, the arterioles branch into thousands of even smaller blood vessels, known as capillaries. The wall of a capillary is only one cell thick and molecules can diffuse across it easily. This means that substances can be exchanged between the blood in the capillaries and the surrounding cells. In the respiring tissues, oxygen and glucose diffuse out of each capillary, while waste products, such as CO_2, are picked up. In the lungs, it is CO_2 that diffuses out of the capillaries and oxygen that is picked up (see unit 18).

- Eventually the capillaries join together again to form venules, and these join together to form the veins that will carry the blood back to the heart. The vein that returns blood to the right side of the heart is known as the **vena cava** and the vein that returns it to the left side is known as the **pulmonary vein**. By the time blood reaches the veins it has travelled a long way from the heart and its pressure has dropped to a very low level. In order to ease the passage of low-pressure blood, veins have a large **lumen**, which is the name we give to the hollow space inside a blood vessel, and **pocket valves** which stop the blood from flowing backwards. A diagram showing how the pressure of the blood changes, from arteries to veins, is shown below.

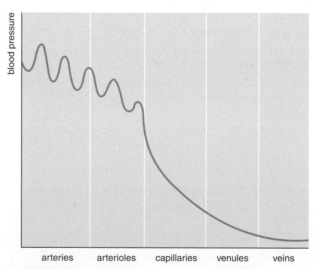

- As you can see in the diagram, the blood pressure in the arteries and arterioles fluctuates up and down. A new pressure peak, or **pulse**, is generated every time the heart contracts and squirts blood out into the vascular system. If you press a finger against the artery in someone's wrist, you can feel the pressure peaks passing by. If the person's heart starts beating faster, then more pressure peaks will be generated every minute and you will be able to feel their **pulse rate** increasing. As the pressure peaks travel away from the heart, they are gradually dampened down by the elasticity of the artery and arteriole walls until eventually, on arrival at the capillaries, the flow of blood is smooth and even.

- The blood itself is made of water, with a variety of substances dissolved in it, and floating blood cells. The watery part of blood, excluding the cells, is known as **plasma**. A chart listing the contents of the plasma is given below.

Picked up from outside the body and carried to the cells	Picked up from inside the body and carried away to be excreted	Carried from one place to another inside the body	Always in the blood
oxygen glucose amino acids lipids minerals vitamins	CO_2 urea	hormones	plasma proteins red blood cells white blood cells

- There is enough room in the plasma to carry, in dissolved form, all the glucose, amino acids, and mineral ions that the body needs. This is not true for oxygen, however. Because the solubility of oxygen in water is very low, only a small amount can be carried in dissolved form. The rest has to be carried by molecules of **haemoglobin**, a protein pigment found in the red blood cells.

- A molecule of haemoglobin consists of four polypeptide subunits (see unit 3), each of which contains a **haem group** that can bind to a single molecule of oxygen. This means that one molecule of haemoglobin can carry four molecules of O_2. In order to function effectively as an oxygen transport molecule, haemoglobin must pick up oxygen in the lungs, carry it round the body, and then drop it off in the respiring tissues where it is needed. In order to do this it must have some way of detecting whether it is in the lungs, where it should be **loading** with oxygen, or in the tissues, where it should be **unloading**. The main clue that haemoglobin responds to is the concentration of oxygen, the pO_2, in the surrounding blood plasma. In the capillaries that run near the lungs, where oxygen is absorbed, there is a high pO_2 and this triggers loading. In the capillaries that run through the respiring tissues, which are continuously using up oxygen, there is a low pO_2 and this triggers unloading. Another clue that haemoglobin responds to is the concentration of CO_2 in the plasma. Respiration produces CO_2, and the high concentration of this gas in the capillaries that run through the respiring tissues causes haemoglobin to unload even more oxygen than usual, a phenomenon known as the **Bohr effect**. The pattern of oxygen loading and unloading that we have just described is illustrated by the **dissociation curves** shown below.

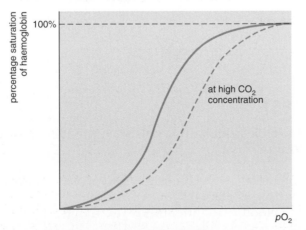

The **percentage saturation of haemoglobin**, on the y axis of the graph, is a measure of how much oxygen the haemoglobin is carrying. At high values of pO_2, which trigger oxygen loading, the haemoglobin has a high percentage saturation. At low values of pO_2, which trigger unloading, it has a low saturation. If a lot of CO_2 is present, the Bohr effect is triggered and the haemoglobin drops off oxygen to become even less saturated than usual, as illustrated by the dotted line.

TESTS

RECALL TEST

1 Arrange the following in the correct order, beginning with capillaries:

arterioles, veins, capillaries, arteries, venules, the heart

_____ (1)

2 Why do arteries have thick muscular walls?

_____ (1)

3 What is the function of the capillaries?

_____

_____ (1)

4 Why do veins contain pocket valves?

_____

_____ (1)

5 Why does blood pressure in the arteries fluctuate?

_____

_____

_____ (1)

6 What is 'blood plasma'?

_____

_____ (1)

7 How many molecules of O_2 can a haemoglobin molecule carry?

_____ (1)

8 What is pO_2 a measure of?

_____ (1)

9 What is the 'Bohr effect'?

_____

_____

_____ (1)

10 What is a lumen?

_____ (1)

(Total 10 marks)

CONCEPT TEST

1 Complete the table below with a tick (✔) if the statement is true or a cross (✗) if it is not true.

	carry oxygenated blood	connect to arterioles	involved in gas exchange	have a large lumen	show non-pulsed blood flow
Arteries					
Capillaries					
Veins					

(3)

2 As people get older, the walls of their arteries become less elastic. What effect do you think this has on their blood pressure?

_____

_____

_____ (2)

3 The graph right shows two dissociation curves: one for human haemoglobin and one for the haemoglobin of the llama, an animal that lives at high altitudes. Explain the difference between the two curves.

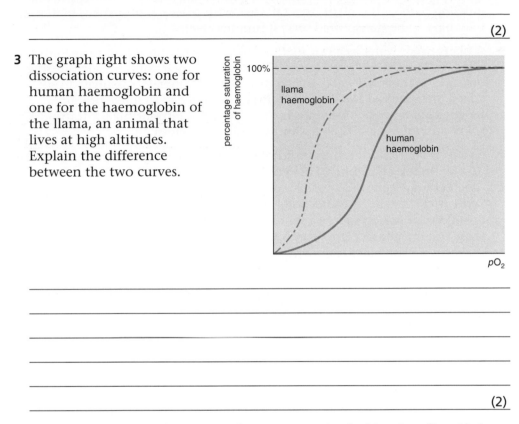

_____

_____

_____

_____

_____ (2)

4 A number of chemical reactions that can occur in the blood are listed below.

$$CO_2 + H_2O \rightarrow H_2CO_3$$
carbonic acid

$$H_2CO_3 \rightarrow H^+ + HCO_3^-$$
hydrogen ion   hydrogen carbonate ion

$$H^+ + HbO_2 \rightarrow HHb + O_2$$
oxyhaemoglobin   haemoglobic acid

Use the information given to explain how the Bohr effect occurs.

_____

_____

_____

_____

_____ (3)

(Total 10 marks)

TRANSPORT IN PLANTS

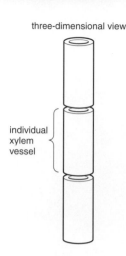

three-dimensional view

individual
xylem
vessel

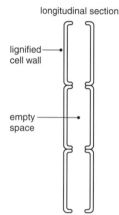

longitudinal section

lignified
cell wall

empty
space

Xylem

- A plant, just like an animal, has to transport a variety of different substances from one place to another inside its body. Water, and the dissolved mineral salts that it contains, must be moved from the roots up to the leaves. Sugar, which has been made in the leaves by photosynthesis, must be carried to the growing parts of the plant. In a tall tree, with a deep root system, materials may have to be transported several hundred feet.

- A plant needs water as a raw material for photosynthesis and to keep its cells filled, or **turgid**. If its cells aren't filled so that their contents are pressing out against their cell walls, the plant will wilt. Minerals, dissolved in soil water, are also vital to the plant. Nitrates are needed to make protein, magnesium to make chlorophyll, etc. The tissue responsible for transporting water and mineral salts in a plant is known as **xylem**.

- Xylem is built out of tall, cylindrical cells, known as **xylem vessels**, stacked up one on top of another. These cells are dead and empty; all that remain are their cellulose cell walls (which are strengthened by a polymer called **lignin**). Soon after they die, the end walls, which separate the vessels in a stack, break down. This produces a set of hollow tubes which run along inside the stem and branches of the plant and out into each leaf, as **veins**. It is up these tubes that water is drawn.

- The continuous flow of water through a plant is known as the **transpiration stream**. It is driven by the **evaporation** of water from the leaves. The underside of a leaf is covered in thousands of tiny pores, called **stomata**. As water molecules evaporate out of the stomata, the liquid inside the surrounding cells becomes more concentrated. This draws water in, by osmosis, from neighbouring cells, the contents of which, in turn, become more concentrated. In this way, water is drawn down a concentration gradient all the way from the xylem vessels, in the veins of the leaf, to the stomata on its underside.

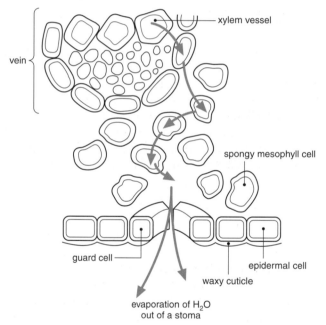

xylem vessel

vein

spongy mesophyll cell

guard cell

epidermal cell

waxy cuticle

evaporation of H_2O
out of a stoma

A section through a leaf. The arrows show the movement of water.

- The movement of water up the xylem itself depends on the fact that water molecules are polar and stick to one another by hydrogen bonding. We call this phenomenon **cohesion**. As osmosis draws water molecules out of the xylem vessels in the leaf, these molecules pull others up behind them so that a linked chain of water molecules is reeled up the xylem. Due to the strength of the evaporative pull, this chain is stretched out, or under **tension**.

- At the base of the xylem, the pull of the chain draws in new water molecules from the cells of the root. This creates a concentration gradient in the root similar to that in the leaf, with water molecules moving, by osmosis, from the soil to the outermost cells of the root and from there to the cells adjacent to the xylem.

- The outermost layer of cells in the root is known as the **epidermis**. Many of the cells in the epidermis have long hair-like extensions. These are known as **root hairs** and they increase the surface area available for the absorption of water.

- As water travels through the root, from the epidermis to the xylem, there are two possible pathways that it can follow. If it travels from cell to cell, crossing a series of cell walls and cell membranes in the process, we say that it is following the **symplastic** pathway. If it travels entirely via cell walls, without crossing any membranes, we say that it is following the **apoplastic** pathway. About halfway into the root, water encounters the **Casparian strip**. This is a band of waxy material, called **suberin**, which runs through the walls of the cells in the middle of the root and is impermeable to water. The Casparian strip blocks the apoplastic pathway and forces water to travel via the symplastic pathway for a bit. This guarantees that the water has been filtered through a membrane before it arrives at the xylem.

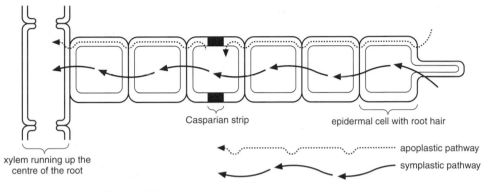

The pathways followed by water as it passes through a root

- As well as water, sugars and other organic molecules, such as amino acids, must also be moved from one place to another inside the plant. These **assimilates** have to be transported from the leaves, where they are made, to the growing parts of the plant, such as the shoots and the roots, where they are needed for energy and growth. The tissue responsible for transporting organic molecules in a plant is known as **phloem**.

- Phloem tissue contains two different types of cell, **sieve-tube cells** and **companion cells**. Sieve-tube cells are tall, cylindrical cells, which are stacked up, one on top of another, rather like xylem vessels. Unlike xylem vessels, however, the sieve-tube cells of phloem are alive and filled with cytoplasm. The end walls that separate one sieve-tube cell from another are not completely broken down but are, instead, perforated by many tiny holes which allow material to move from one cell to the next. These perforated end walls are known as **sieve plates**. Adjacent to the sieve-tube cells are the companion cells, which are also alive and filled with mitochondria.

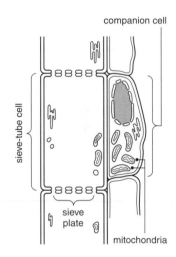

Phloem (longitudinal section)

- The movement of organic molecules through the sieve tubes of a plant is known as **translocation**. Unlike transpiration, this process requires energy. At the end of the sieve tube where the assimilates are being produced, which is called the **source**, energy is used to move organic molecules into the phloem. This process is known as **active loading**. At the end of the sieve tube where the assimilates are going, which is known as the **sink**, energy is used to move organic molecules out of the phloem. This means that the source end of the sieve tube is

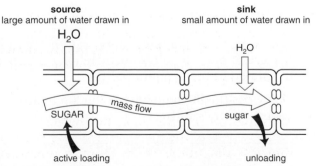
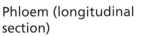

more concentrated than the sink end and, as a result, it draws in more water, by osmosis, than the sink end does. This difference in water uptake creates a pressure difference which forces liquid along the sieve tube, from source to sink, in a process known as **mass flow**.

TESTS

RECALL TEST

1 Distinguish between 'transpiration' and 'translocation'.

_____

_____ (1)

2 What force drives the transpiration stream?

_____ (1)

3 What feature of root epidermal cells helps to maximize the absorption of water?

_____

_____ (1)

4 Distinguish between the apoplastic and symplastic pathways of water movement.

_____

_____

_____ (1)

5 What is the Casparian strip made of?

_____ (1)

6 What is a sieve plate?

_____ (1)

7 What feature of a companion cell supports the idea that translocation is an active process?

_____ (1)

8 What is meant by 'source' and 'sink'?

_____

_____ (1)

9 What is 'active loading'?

_____

_____ (1)

10 What do we call the bulk movement of liquid from source to sink?

_____ (1)

(Total 10 marks)

CONCEPT TEST

1 When the stem of a transpiring plant is cut in half with scissors, the water in the xylem of each half immediately pulls back from the cut point. This is illustrated in the diagram right.

How does the movement of water after cutting provide evidence for the cohesion–tension theory of transpiration?

_____

_____

_____

_____ (3)

stem cut here

H_2O

H_2O

end of stem sealed

84

2 The graph below shows how the branch of a small apple tree varied in diameter over a 24-hour period.

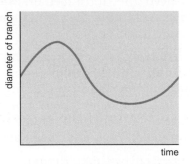

a Draw a curve on the blank graph below to show the rate of transpiration of the apple tree varied over the same 24-hour period.

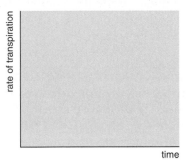

(2)

b How does the change in diameter of the branch provide evidence for the cohesion–tension theory of transpiration?

_____

_____

_____

_____

_____

_____ (3)

3 Complete the table below with a tick (✔) if the statement is true or a cross (✗) if it is not true.

	contains lignified cells	contains cells with living contents	contains cells with incomplete end walls	transport is inhibited by metabolic poisons	transports the products of photosynthesis
Xylem					
Phloem					

(2)

(Total 10 marks)

HETEROTROPHIC NUTRITION

- **Heterotrophs** are organisms that, unlike plants, cannot manufacture their own organic molecules from simple inorganic ones. Instead, to obtain the organic molecules they need, they must feed on other organisms. They are, in effect, nutritional thieves, stealing the components of other creatures' bodies in order to construct their own.

- Human beings are **omnivores**, a type of heterotroph that preys on a wide range of other organisms, including both plants and animals. A balanced human diet must contain all the major types of organic molecule. **Carbohydrates** (see unit 2) can be obtained from foods such as bread, rice, and potatoes. **Lipids** (see unit 2) are found in fatty or oily foods. **Proteins** (see unit 3) are present in meat, eggs, and vegetables such as the soya bean.

- **Vitamins** are organic molecules which, though they are only needed in small quantities, are also a necessary part of the human diet. An example of a vitamin is **nicotinic acid**, or **B$_3$**. This can be obtained from wholemeal bread, yeast extract, and meat, particularly liver. It is used in the construction of **NAD**, a hydrogen-carrying molecule which plays a vital role in respiration (see unit 8). A person whose diet is lacking in nicotinic acid is likely to suffer from skin rashes and lesions, a condition known as **pellagra**.

- As well as all these organic nutrients, a balanced diet must also contain a range of simple inorganic molecules, or **mineral ions**. Examples include **calcium**, which is needed for the construction of bones, **iron**, which forms a vital part of haemoglobin (see unit 20), and **phosphate**, which is needed in the manufacture of a whole range of molecules, from phospholipids (see unit 6) to ATP (see unit 8).

- All heterotrophs, whatever their dietary requirements, suffer a common problem. Most molecules of food are too big to pass easily across an organism's surface, from the world outside to the cells inside its body. The solution to this problem is **digestion**, a process in which food molecules are broken down into small, soluble sub-components. Once food has been digested, the soluble molecules that result can diffuse across the organism's surface and into its body, a process known as **absorption**.

- In humans, as in most animals, the site of digestion and absorption is the **gut**, or **alimentary canal**. A diagram of the human alimentary canal is shown on the left.

- Food begins its journey in the mouth, where chewing, or **mastication**, occurs. This process breaks the food up into small, manageable pieces, which not only makes swallowing easier, but also increases the surface area of the food. This increase in surface area means that a digestive enzyme, called **amylase**, which is present in saliva, can work on the food more effectively.

- Like all the digestive enzymes in the gut, amylase is a **hydrolytic** enzyme. It catalyses a reaction in which water is added to split apart bonds formed by condensation (see unit 1). Different hydrolytic enzymes are needed to break bonds in different types of food molecule. Amylase is used to break **starch** down into **maltose** (see unit 2).

- After chewing, the food is swallowed and passed into the **oesophagus**, a long, muscular tube which runs down to the stomach. Waves of muscular contraction, in the walls of the oesophagus, squeeze the food along in a process known as **peristalsis**.

- Food may remain in the stomach for up to four hours, while the enzymes contained in the acidic **gastric juice** break its components down. After this it is released into the **duodenum**, where yet more enzymes get to work on it. Some of these enzymes are attached to the duodenal lining, and some arrive down the **pancreatic duct**, in **pancreatic juice**. A summary of all these enzymes is given above right.

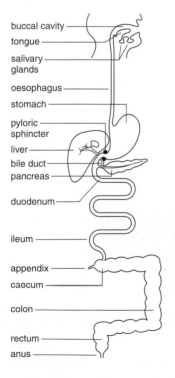

buccal cavity
tongue
salivary glands
oesophagus
stomach
pyloric sphincter
liver
bile duct
pancreas
duodenum
ileum
appendix
caecum
colon
rectum
anus

Site of production	Enzyme	Site of action	Substrate	Product
salivary glands	amylase	mouth	starch	maltose
stomach lining	pepsin	stomach	proteins	peptides
pancreas	amylase	duodenum	starch	maltose
	trypsin	duodenum	proteins	peptides
	lipase	duodenum	lipids	fatty acids + glycerol
lining of duodenum	disaccharidases	duodenum	disaccharides	monosaccharides
	exopeptidases	duodenum	peptides	amino acids
liver	bile salts (not an enzyme)	duodenum	lipids	lipid droplets

Protein-digesting enzymes, such as **pepsin**, are released in an inactive form. This is to prevent them from breaking down the cells where they are made. Once released, they can be activated. The only substance on the list which isn't an enzyme is **bile**. This doesn't actually hydrolyse fats, it just **emulsifies** them, causing them to break up into little droplets, which increases their surface area and allows enzymes to work on them more effectively.

● After digestion in the duodenum, food is passed on to the **ileum**, where it is absorbed. A diagram of the ileum lining is shown below.

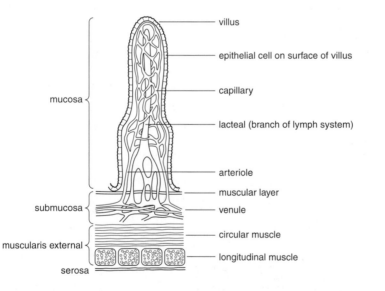

As you can see in the diagram, the ileum lining is folded to form finger-like processes called **villi**. The cell membranes of the **epithelial cells**, which are found on the surface of these villi, are also folded, to form **microvilli**. All this folding dramatically increases the total surface area available for absorption.

● Almost all of absorption is due to **diffusion**, though a little **active transport** may be involved as well (see unit 6). Sugars and amino acids diffuse into the blood capillaries in each villus. Fatty acids and glycerol enter the **lacteals**, which are branches of the **lymph system**, from where they will, eventually, make it into the blood.

● All the indigestible components of the food, which are too large to be diffuse through the gut wall, carry on into the **colon,** where water is reabsorbed, and then out of the **rectum** as **faeces**.

TESTS

RECALL TEST

1 What is a 'heterotroph'?

_____

_____ (1)

2 What is an 'omnivore'?

_____

_____ (1)

3 What deficiency disease results from a lack of vitamin B_3?

_____ (1)

4 Name *two* mineral ions which are necessary in a balanced human diet.

_____ (1)

5 How does mastication help enzymes to work more efficiently?

_____

_____ (1)

6 What, exactly, does a hydrolytic enzyme do?

_____

_____ (1)

7 What is 'peristalsis'?

_____

_____ (1)

8 Why are protein-digesting enzymes released in an inactive form?

_____

_____ (1)

9 Name three enzymes found in the pancreatic juice.

_____

_____ (1)

10 What occurs in the colon?

_____ (1)

(Total 10 marks)

CONCEPT TEST

1 The graph below shows the pH (acidity/alkalinity) changes that result when certain substances are combined together.

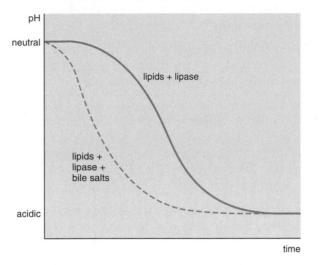

a Why does a combination of lipids and lipase result in a pH decrease?

_____

_____

_____ (1)

b Why does the addition of bile salts make this pH decrease happen faster?

_____

_____

_____

_____ (2)

3 Pepsin, which is found in the stomach, is an endopeptidase enzyme. It attacks the peptide bonds in the middle of a polypeptide chain, breaking the chain up into shorter lengths. The protease enzymes attached to the lining of the ileum are exopeptidase enzymes. These start at one end of a polypeptide chain and work their way along the chain, breaking off one amino acid at a time.

Explain the benefit that is gained from exposing proteins to an endopeptidase before they are acted on by exopeptidases.

_____

_____

_____

_____ (3)

4 The diagram right shows a columnar epithelial cell, one of the cells found lining the ileum.

Describe *two* features of this cell that help it absorb material effectively.

a _____

_____ (2)

b _____

_____ (2)

(Total 10 marks)

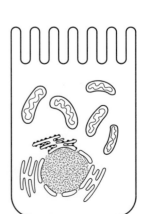

REPRODUCTION IN HUMANS

- The first living organisms reproduced **asexually**, by simple splitting, or budding. This is a straightforward and efficient way to make more organisms. The only problem is that, unless a **mutation** occurs (see unit 16), the offspring of asexual reproduction are all genetically identical. **Sexual reproduction** evolved as a way to produce offspring that would show **genetic variation**.

- The benefit of producing offspring that differ from one another is that, in a variable or rapidly changing environment, at least some of these offspring will have characteristics that allow them to survive.

- Sexual reproduction requires two parents. Each of these parents produces a **haploid** cell, known as a **gamete**, which contains half the normal quantity of DNA. Gametes are produced by a type of cell division known as meiosis (see unit 14). When the gametes of the two parents fuse, in a process known as **fertilization**, a **zygote** is produced. This zygote will develop into a new, and unique, organism that contains a mixture of DNA from both parents.

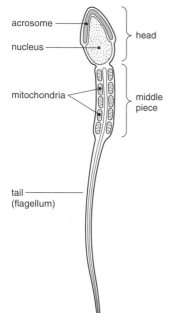

acrosome

nucleus

head

mitochondria

middle piece

tail (flagellum)

- In most animal species, there are two distinct types of gamete producer: male and female. Males produce large quantities of small gametes called **spermatozoa**, or sperm, for short. Females produce a limited number of large gametes, each of which is known as an **ovum**, or **egg**. In mammals, such as human beings, the fusion of sperm and egg occurs inside the female, and it is inside the female that the zygote develops into a new organism, a baby.

- Human males begin producing sperm at **puberty**, the beginning of the teenage years. Sperm are produced in the **testes**, which hang outside the body in the **scrotal sac**. This location keeps them at a temperature which is slightly lower than the rest of the body, and ideal for sperm production. Ten million sperm are produced by every gram of testis, every day. The structure of a human sperm cell is shown left.

- When a man **ejaculates**, during sex, sperm are released out of his **penis** and into the woman's **vagina**. If fertilization is to occur, the sperm must swim up from the vagina, through the woman's womb and into her **fallopian tubes**, where, with luck, they may encounter an egg. A sperm cell swims by beating its long tail from side to side. The energy to drive the beating of the tail is provided by mitochondria (see unit 5) in the middle piece of the sperm cell. The **acrosome**, located in the head of the sperm cell, is a vesicle, or tiny sac, that contains digestive enzymes. When a sperm cell encounters an egg, its acrosome breaks open and the enzymes that are released out of it digest a pathway through the outer layer of the egg, allowing the sperm cell to enter.

- The eggs, or ova, of human females are produced in the **ovaries**. One egg is produced every month, starting at puberty and carrying on until the onset of **menopause**, the post-reproductive phase of a woman's life which begins when she is in her late forties or early fifties. The series of cell divisions involved in egg production is shown here.

- As you can see, the meiotic divisions involved in egg production are uneven. In each division a tiny **polar body**, containing half the divided DNA, is discarded. This means that you end up getting one big gamete instead of four small ones.

- Once a month, as well as producing an egg, a woman's body also begins to prepare for the possible fertilization of the egg. This

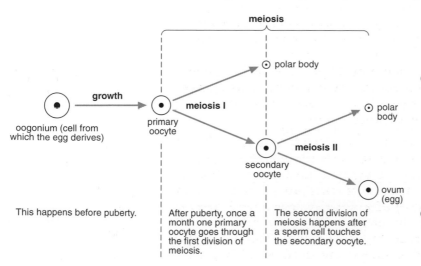

meiosis

polar body

growth

oogonium (cell from which the egg derives)

primary oocyte

meiosis I

polar body

secondary oocyte

meiosis II

polar body

ovum (egg)

This happens before puberty.

After puberty, once a month one primary oocyte goes through the first division of meiosis.

The second division of meiosis happens after a sperm cell touches the secondary oocyte.

preparation involves the growth and thickening of the womb lining or **endometrium**. If the egg is fertilized, the zygote that results will grow by mitotic division to become a little ball of cells called a **blastocyst**. The thickened womb lining provides a place for the blastocyst to attach to. Once attached, or **implanted**, the blastocyst can start drawing nutrients from its mother and developing into a baby. If the egg isn't fertilized, the blood-filled endometrium is shed at the end of the month in a process known as **menstruation** and a new womb lining begins to develop.

● The monthly cycle of egg production and endometrium growth in a woman is known as the **menstrual cycle**. This cycle is controlled by a set of four hormones. Two of these hormones, **follicle-stimulating hormone (FSH)** and **luteinizing hormone (LH)** are produced by a gland in the brain called the **pituitary gland**, from where they travel, via the bloodstream, to the ovaries and womb. The other two hormones, **oestrogen** and **progesterone**, are produced in the ovaries themselves. A graph showing the changing levels and effects of these hormones, throughout the menstrual cycle, is shown below.

● The cycle begins with the release of FSH. This hormone triggers the growth of a single **follicle** in one of the ovaries. A follicle is a small ball of cells that contains a primary oocyte. As the follicle grows the oocyte inside starts to develop, passing through all the stages of the first meiotic division to become a secondary oocyte. The growing follicle also begins to release the hormone oestrogen, which causes the endometrium to start thickening. On the graph you can see that a peak in the level of FSH is followed by a peak in the level of oestrogen. After about fourteen days, the high level of oestrogen in

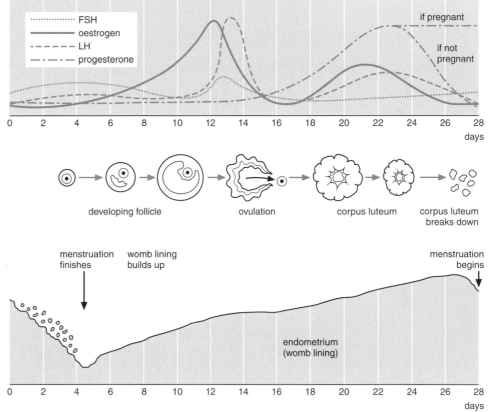

the bloodstream causes the pituitary gland to stop producing FSH and to start producing LH. LH triggers the process of **ovulation**, in which the follicle breaks open and the secondary oocyte is released out of the ovary and into one of the fallopian tubes that lead down to the womb. After it has released the oocyte, the ruptured follicle turns into a structure called the **corpus luteum**. The corpus luteum then starts to produce the hormone progesterone, which is needed to maintain the uterine lining in its thickened state. On the graph you can see that a peak in the level of LH, at fourteen days, is followed by a peak in the level of progesterone.

● If fertilization fails to occur, the corpus luteum eventually breaks down and ceases to produce progesterone. Without progesterone to maintain it, the uterine lining is then shed and the whole cycle starts up again. If fertilization does occur, the implanted blastocyst that results starts producing a hormone called **human chorionic gonadotrophin**, or **HCG**, which it releases into the mother's bloodstream. HCG helps to maintain the corpus luteum in the mother's ovary, so that it keeps on producing progesterone. This means that the uterine lining, and the implanted blastocyst, are saved from being shed and pregnancy can progress.

TESTS

RECALL TEST

1 What cell is produced when two gametes fuse together?

_____ (1)

2 Where does gamete fusion typically occur?

_____ (1)

3 Why does the scrotal sac hang outside the body?

_____
_____ (1)

4 What is the name of the large vesicle, containing digestive enzymes, that is found inside the head of a sperm cell?

_____ (1)

5 What is 'menopause'?

_____
_____ (1)

6 What is a polar body?

_____
_____ (1)

7 What is the function of the endometrium?

_____
_____ (1)

8 What happens during menstruation?

_____
_____ (1)

9 How many days into the menstrual cycle does ovulation occur?

_____ (1)

10 What is the function of the hormone HCG?

_____
_____
_____ (1)

(Total 10 marks)

CONCEPT TEST

1 Aphids, or greenfly as they are more commonly known, are capable of both asexual and sexual reproduction. In the summer months, when food is readily available, an aphid population consists entirely of asexually reproducing females which give birth to new females. Just before the onset of the harsh winter months, a few males are born and sexual reproduction occurs.

Explain why asexual reproduction happens in the summer and sexual reproduction just before winter.

_____

_____

_____

_____

_____

(3)

2 Fill in the empty boxes in the table given below.

Hormone	progesterone	LH		
Site of production		produced by pituitary gland		produced by follicle
Effect	maintains uterine lining		triggers follicle development	

(3)

3 The 'combined pill' is an oral contraceptive that contains high levels of both oestrogen and progesterone. One pill is taken every day for 21 days, and then seven days are allowed to pass without taking the pill, before the cycle is started again.

a Using your knowledge of reproductive hormones, explain how the combined pill prevents pregnancy from occurring.

_____

_____

_____

_____

(2)

b Suggest why it is necessary to come off the pill for one week out of every four.

_____

_____

_____

(2)

(Total 10 marks)

REPRODUCTION IN FLOWERING PLANTS

- A tulip, a chestnut tree, and a stalk of wheat may appear to be very different types of plant, but there is one important thing that they have in common: they all possess reproductive organs known as **flowers**. The flowers of the wheat plant may be less obvious than those of the tulip or the chestnut tree, because they lack brightly coloured petals and a pleasant smell, but they are flowers nevertheless. All three species belong to a group of plants known as the **angiosperms**, or flowering plants (see unit 17).

- Flowers are described as reproductive organs because they are the place where male and female **gametes** are produced. Gametes are **haploid** cells which are produced by **meiosis** (see unit 14). Sexual reproduction in plants involves the fusion of a male gamete from one plant with a female gamete from another plant. The result of this fusion is a **zygote**, a **diploid** cell which will develop into a new and genetically unique individual.

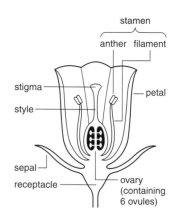

- As humans, we are used to the idea that individuals are either males, who produce male gametes, or females, who produce female gametes. In flowering plants, things can get a bit more complicated. Though there are some plant species, such as yew and holly, that have separate sexes, most angiosperms have **hermaphrodite** flowers, which can produce both male and female gametes. While the male gametes are being sent off to fertilize the female gametes of other plants, the flower's own female gametes are sitting there waiting to be fertilized themselves. A diagram of a hermaphrodite flower is shown on the left.

- Female gametes are produced inside the **ovary** of the flower, where they remain until they are fertilized. In some types of flower, the ovary may hold as many as a hundred female gametes, each one contained inside a separate **ovule**. The flower shown in the diagram has just six ovules. The series of cell divisions that leads to the production of a female gamete, inside an ovule, is shown below.

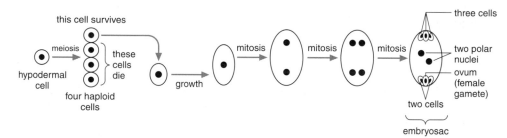

- The sequence begins with a **hypodermal cell**, which is just one of the normal, diploid cells that make up the ovule. After one meiotic division (see unit 14) and three mitotic divisions (see unit 13), this produces a structure called the **embryosac** which contains eight haploid nuclei, one of which is the **ovum**, or female gamete. By the end of the maturation process, every ovule in the ovary of the flower will contain a single embryosac with a single ovum, or female gamete, inside it.

- The production of male gametes, in the **anthers** of the flower, involves a slightly different series of cell divisions.

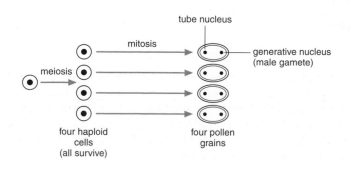

- As before, the sequence starts with a hypodermal cell, which is just one of the normal, diploid cells that make up the anther. This hypodermal cell undergoes a single meiotic division to produce four haploid daughter cells. In male gamete production, in contrast to female, all of these daughter cells survive and each one undergoes a further mitotic division to produce a **pollen grain** containing two identical haploid nuclei, one of which, the **generative nucleus**, is the male gamete. It is important to note that the pollen grain is the structure that contains the male gamete, in the same way that the embryosac contains the female gamete. *The pollen grain is not, itself, a gamete.*

- The function of the pollen grain is to carry the male gamete from the flower where it was produced to the flower of another plant, where it will be able to fertilize a female gamete. The transfer of pollen from the anther of one flower to the **stigma** of another is known as **pollination**. The flower shown in the diagram on the previous page is an **insect-pollinated** flower. It has brightly coloured petals and nectar to attract insects. When an insect visits the flower it gets dusted with sticky pollen which will then get rubbed off on the stigma of the next flower it visits. Other flowers, such as those found on a wheat plant, are **wind-pollinated**. They produce large quantities of very light pollen grains which are carried away by the wind and picked up by the dangling, feathery stigmas of other wind-pollinated flowers.

- Once a pollen grain has landed on the stigma of a compatible flower, it **germinates** and produces a **pollen tube** which begins to grow down through the **style** of the flower towards the ovary. The growth of the pollen tube is controlled by the **tube nucleus**. The generative nucleus, or male gamete, travels down the pollen tube just behind the tube nucleus and in this way it is able to get down to an ovule, inside the ovary of the flower, where it can fuse with a female gamete.

- Once the pollen tube has reached an ovule, it stops growing, and the tube nucleus, having done its job, breaks down. The generative nucleus then undergoes a final mitotic division and splits in half to produce **two** male nuclei. One male nucleus fuses with the ovum, or female gamete, at the bottom of the embryosac, to produce a diploid zygote. The other male nucleus fuses with the two **polar nuclei**, in the middle of the embryosac, to produce a **triploid endosperm nucleus**. This **double fertilization** is unique to flowering plants.

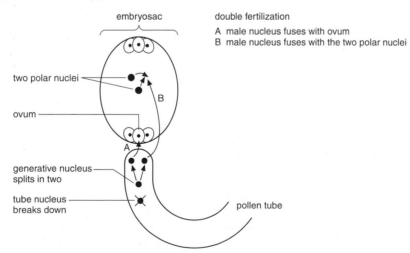

- After fertilization, the ovule develops into a **seed**. The zygote grows to become the **embryo**, or baby plant, inside the seed, while the triploid endosperm nucleus divides away repeatedly to produce the seed's **endosperm food store**. The integuments, which once surrounded the ovule, become the seed's coat or **testa** and the ovary, in which the seed is located, swells up to become a **fruit**.

TESTS

RECALL TEST

1 What is the official scientific name for flowering plants?

_____ (1)

2 What is a hermaphrodite flower?

_____

_____ (1)

3 What does a stamen consist of?

_____ (1)

4 How many nuclei does an embryosac contain?

_____ (1)

5 Where are pollen grains produced?

_____ (1)

6 Why do wind-pollinated plants have a feathery stigma?

_____

_____ (1)

7 What is the function of the tube nucleus?

_____

_____ (1)

8 Why is the fertilization that happens in flowering plants described as a 'double fertilization'?

_____

_____

_____

_____ (1)

9 What does the seed coat, or testa, develop from?

_____ (1)

10 What does the ovary develop into after fertilization?

_____ (1)

(Total 10 marks)

CONCEPT TEST

1 Complete the table below with a tick (✔) if the statement is true or a cross (✗) if it is not true.

	found in ovule	is the male gamete	produces a pollen tube	produced by meiosis and mitosis	stays in the flower
Embryo-sac					
Pollen grain					

(2)

2 The graph below shows how the activity inside a tube nucleus and a generative nucleus vary during the growth of a pollen tube.

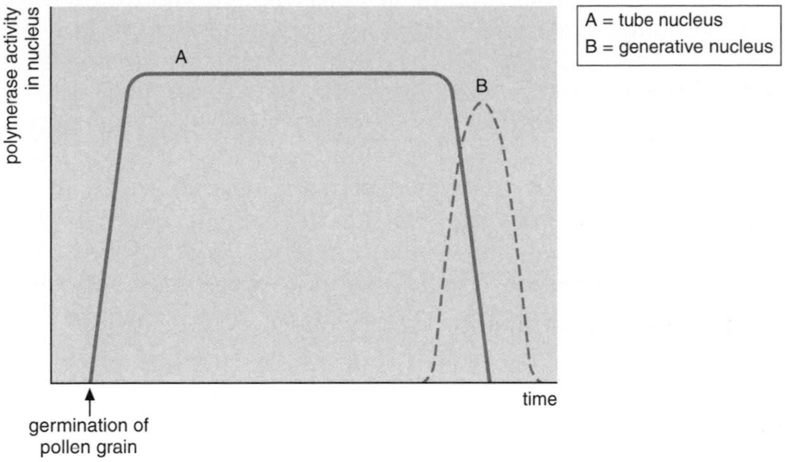

a Explain the shape of curve A.

_____

_____

_____

_____ (2)

b Explain the shape of curve B.

_____

_____

_____

_____ (2)

3 A pollen grain from a plant with genotype AA lands on the stigma of a plant with genotype aa and fertilization results. Work out the genotype of the following structures.

a The generative nucleus (male gamete).

_____ (1)

b The zygote formed by the fertilization.

_____ (1)

c The endosperm food store of the seed that develops.

_____ (1)

d The fruit that the seed is contained in.

_____ (1)

(Total 10 marks)

AN INTRODUCTION TO ECOLOGY

● There are nine planets orbiting the Sun and many of these have orbiting moons. Some of these worlds are freezing cold and some of them boiling hot. On Venus, the surface temperature is high enough to melt lead. Across the Solar System there are atmospheres of hydrogen, of carbon dioxide, and of methane. There are worlds wracked by volcanic activity or by storms that last for a thousand years. There are worlds where sulphuric acid falls as rain. On the third planet out from the Sun, however, there is an even stranger phenomenon. Here, in a thin layer clinging to the surface of the planet, there is liquid water and an oxygen atmosphere. Within this layer, which is known as the **biosphere**, three million different species of living organism breed and feed and compete with one another. **Ecology** is the science that tries to make sense of what is going on in the biosphere. It is the study of living organisms and the way that they interact with their **environment**.

● An organism's environment consists of everything that surrounds it. This includes all the other living creatures that it interacts with, its **biotic environment**, as well as all the non-living components of the world, such as water, soil, and the gases in the atmosphere, which make up its **abiotic environment**. Every day, a living organism has hundreds of interactions with its environment. When it eats food, in the form of another organism, or when it catches a disease, or when it mates or competes for territory with its fellows, it is interacting with its biotic environment. When it drinks a mouthful of water, or breathes in oxygen and breathes out CO_2, or urinates onto the ground, it is interacting with its abiotic environment.

● An **ecosystem** is a definable area containing a self-sustaining collection of living creatures interacting with each other and with the abiotic environment. A pond, for example, can be viewed as an ecosystem. To start with, it is a neatly definable area, since it consists of water surrounded by land. It contains a collection of living organisms of different species, a **community**, including algae, water lilies, fish, water beetles, pond snails, etc., and all of these interact with each other and the abiotic environment of the pond. The fish eat the beetles and the beetles eat the algae. Gases and nutrients are released and taken up. Finally, the pond community is more or less self-sustaining. The oxygen used up by the animals is replaced by algal photosynthesis. The nutrients used up by the algae are replaced when the animals die and decay. Unless there is a major environmental change, such as a drought, the pond will go on being a pond, with new organisms born to replace those that have died, for generation after generation.

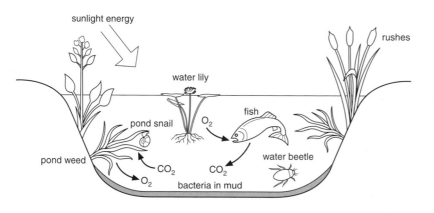

A pond ecosystem

● A fish tank, unlike a pond, is not an ecosystem. This is because it is not self-sustaining. If you stop feeding the fish, or switch off the pump that oxygenates the water, the network of interactions within the tank will soon become unbalanced and you will probably end up with a load of dead fish. Why is the pond stable, when the tank isn't? The answer has to do with size. The larger size of the pond allows it to support a greater number and diversity of organisms and it is this that gives it stability. If, in the pond, there are too many beetles eating the pond weed, there will be time for the fish population to increase, and eat up the excess of beetles, before the pond weed entirely runs out. If, in the tank, there are too many beetles, the pond weed will be gone in no time. Without any algae left to feed on, the beetles will die and then, without any beetles to eat, the fish will die.

● A community of organisms, within an ecosystem, is made up of a number of different **populations**. A population is a group of organisms of the same species which live in a particular place at a particular time. The pond may contain a population of mayfly larvae, a population of pond snails, a population of moorhens, etc. The place where these populations live, in this case the pond, is described as their **habitat**. Some of the populations may spend their lives within an even more limited region, a **microhabitat**, such as the underside of the stones on the pond bed or in the stems and roots of pond vegetation.

- As well as having a habitat, and sometimes a microhabitat, every population has a particular **ecological niche**. This niche is defined by the way that the population exploits, or makes a living from, its environment. This includes not only where it lives, but also what it feeds on and how and when it feeds on it. The niche of a trout population, within the pond, could be defined as 'pond dwelling, open-water feeding, tadpole and aquatic insect eating'. The niche of a tench population, a different kind of fish, would be 'pond dwelling, bottom feeding, vegetation and detritus eating'. Though both fish populations occupy the same pond, and swim in the same water, they will not be in competition with each other because they occupy different niches and make a living from the pond environment in different ways.

- The feeding, or **trophic**, relationships within a community can be illustrated by drawing a **food web**, which shows which populations feed on which. A food web for a pond ecosystem is shown below:

- As you can see, the arrows in the food web point from populations that are being eaten to the populations that are doing the eating. The arrows are arranged like this in order to

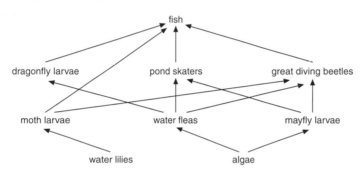

illustrate the direction in which energy flows through the ecosystem, once it has arrived as sunlight. **Energy flow** through an ecosystem is considered in more detail in the next unit, unit 26.

- As well as eating and being eaten by one another, the organisms that make up a community also compete. **Intraspecific competition**, between members of the same species, for food, mates, territory, etc., is the force that drives **natural selection** and the gradual **evolution** of the species (see unit 16). **Interspecific competition**, between individuals of different species, is the force that drives each population into a separate niche, so that competition is reduced. In contrast to all these competitive interactions, some species have a **mutualistic** relationship, in which they help each other out. An example of this is the relationship between nitrogen-fixing bacteria and the leguminous plants (such as peas or beans) that they live inside (see unit 27).

- The populations that make up a community also interact with their abiotic environment. There is a continuous exchange of gases and other nutrients between the pond community and the mud and water of the pond itself. This recycling of nutrients must be exactly balanced if the ecosystem is to remain stable. **Nutrient cycles** are considered in more detail in unit 27.

- Finally, it is worth noting that ecosystems do change over time. A pond may gradually fill up, over hundreds of years, as detritus accumulates on its bed. As the change from pond to swamp occurs, different, better adapted, populations of plants and animals will move in to replace those that went before, in a process known as **ecological succession**. Even more dramatic change may be triggered by human intervention. The pond may be drained, for example, or become polluted. **Human impact** on the biosphere is considered in unit 28.

TESTS

RECALL TEST

1 What do ecologists study?

_____

_____ (1)

2 List three components of your abiotic environment.

_____

_____ (1)

3 What is an 'ecosystem'?

_____

_____

_____ (1)

4 Why is a fish tank not an ecosystem?

_____ (1)

5 What is a 'population'?

_____

_____ (1)

6 What is the 'niche' of an organism?

_____

_____ (1)

7 In what form does energy enter an ecosystem?

_____ (1)

8 What does intraspecific competition result in?

_____ (1)

9 Give an example of a mutualistic relationship.

_____

_____ (1)

10 As a habitat changes, new plants and animals will move in to replace the previous occupants. What do we call this process?

_____ (1)

(Total 10 marks)

CONCEPT TEST

1 The pitcher plant is an insect-eating plant with deep cup-shaped leaves that are filled with liquid. When an insect lands on the rim of a leaf it will often slip in and drown in the liquid. Digestive enzymes, secreted by the leaf, will then break the dead insect down so that the pitcher plant can absorb its nutrients. The leaves of a pitcher plant provide a food-rich environment for a number of different creatures. As well as microorganisms, which feed on drowned insects, there are populations of mosquito larvae and fly larvae which feed on the microorganisms. There are also parasitic wasps that insert their eggs into the bodies of the fly larvae so that their young will have something to feed on when they hatch.

a Within the leaf of a pitcher plant there is a 'community'. What does this 'community' consist of?

_____

_____

_____

_____ (2)

b Draw a food web to show the feeding relationships that occur within the leaf of a pitcher plant.

(2)

c What is the habitat of the fly larvae that are mentioned in the passage?

_____ (1)

d What is the niche of the fly larvae that are mentioned in the passage?

_____

_____ (1)

e Give an example of interspecific competition that occurs within the leaf of a pitcher plant.

_____

_____

_____ (2)

f Give an example of intraspecific competition that occurs within the leaf of a pitcher plant.

_____

_____

_____ (2)

(Total 10 marks)

ENERGY FLOW

- All living organisms are built out of complex organic molecules such as carbohydrates, lipids, and proteins (see units 2 and 3). These molecules contain **energy**. Consider a molecule of glucose inside the leaf of a plant. When a rabbit eats the plant, the molecule of glucose becomes part of the rabbit. When a fox eats the rabbit, the glucose molecule becomes part of the fox. As the glucose molecule moves from organism to organism, so does the energy inside it. We say that there is an **energy flow** through the ecosystem.

- It is photosynthetic organisms, such as plants, that make all the complex organic molecules in the first place. Because of this, we call them **primary producers**.

- A plant makes organic molecules by trapping sunlight energy and using it to stick simple, inorganic, molecules together (see unit 9). The energy that was once in the sunlight ends up inside these organic molecules. *The Sun is the original source of all the energy that flows through the ecosystem.*

- Not all of the incident sunlight energy (i.e. the energy that falls on a plant) is successfully trapped. Some is reflected off the waxy cuticle of the leaf. Some is transferred to the water inside the leaf, causing it to heat up and evaporate. Some sunlight energy may be of the wrong wavelength to be absorbed by chlorophyll. Overall a plant only traps about 1% of the sunlight energy that falls on it. The energy that it does trap and that ends up inside the glucose molecules that have been produced by photosynthesis is called the **gross primary productivity** of the plant.

- There is a variety of things that a plant can do with the glucose molecules that it has made in photosynthesis. Some will be broken down in respiration to release the energy inside them. This energy can be used by the plant to drive vital processes, after which it will be lost out of the plant as heat. Other glucose molecules can be combined together or modified, in chemical reactions, to produce all the other complex molecules that make up the body of the plant. The energy inside these molecules, which is available to an organism that eats the plant, is called the **net primary productivity**.

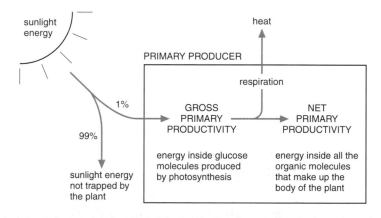

- There are two types of heterotrophic organism (see unit 22) that feed on primary producers. Heterotrophs such as rabbits, which eat plant material when it is still alive, are known as **primary consumers**. Heterotrophs such as fungi and bacteria, which eat plant material after it has died, are known as **decomposers**. When something **decays** it is just being eaten by decomposers, such as bacteria, which are far too small to see.

- The proportion of plant material that is eaten by consumers, and the proportion that is eaten by decomposers, varies from ecosystem to ecosystem. In a grassland ecosystem, such as the African veldt, most of the energy that was trapped by the grass is passed on to grazing consumers such as gazelles, zebras, etc. In a woodland ecosystem, such as an English forest, most of the energy that was trapped by the trees is passed on to decomposers that feed on fallen leaves.

- Even after a mouthful of plant material has been eaten by a grazing consumer, some of the energy in it will still end up being passed to decomposers. This is because a lot of plant matter is tough and indigestible. The energy in the indigestible part of the plant matter is lost to the consumer during **egestion**, when it is passed out of the consumer's gut inside its faeces (see unit 22). Decomposers can obtain energy by feeding on these faeces.

- Once the organic molecules in the digestible part of the plant material have been absorbed into the body of the consumer, there are two things that can happen to them. Some will be broken down in respiration, in which case the energy inside them will be released to drive vital processes and then lost out of the consumer's body as heat. Some of them will be used to build the primary consumer's body. The energy contained in these molecules, which become a permanent part of the consumer, is known as **primary consumer productivity**. In most ecosystems, about 10% of net primary productivity is converted to primary consumer productivity. The remaining 90% is lost to decomposer feeding, in egestion, and as a result of respiration.

- Primary producers, and the primary consumers that feed on them, are only the first two **trophic levels** in a whole **food chain**. This chain carries on to include **secondary consumers**, which feed on the primary consumers, **tertiary consumers**, which feed on the secondary consumers, and sometimes even **quaternary consumers**, which feed on the tertiary consumers. At each stage in this chain, as energy is transferred from one trophic level to the next, a significant proportion is lost to decomposers, in egestion, and as a result of respiration. Gradually, the energy runs out, until there is not enough left to support a further trophic level. This is why food chains are rarely more than four or five links long. Gazelles feed on grass, and lions on gazelles, but there are no giant predators that make a living by feeding on lions. The energy flow through a whole food chain is shown below.

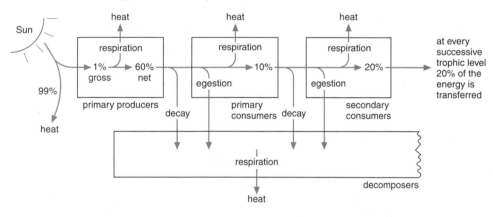

- If you look at the numbers in the diagram you will see that the **efficiency of energy conversion** between primary consumers and secondary consumers is 20%, which is greater than the 10% efficiency between producers and primary consumers. This is because animal tissue is more easily digestible by animals than plant tissue and less energy is lost to egestion.

- In order to work out the efficiency of energy conversion between the trophic levels in a particular ecosystem, ecologists sample animals and plants from the ecosystem and then estimate the total **biomass** of organisms in each trophic level. This data can then be presented in the form of a **pyramid of biomass**; a series of stacked blocks, with each block representing the biomass of a different trophic level. Because of the energy that is lost between trophic levels, each successively higher level in the pyramid has a successively smaller biomass.

secondary consumer biomass

primary consumer biomass

primary producer biomass

TESTS

RECALL TEST

1 Why are plants described as primary producers?

_____

_____ (1)

2 Approximately what percentage of incident sunlight energy is trapped by a photosynthesising plant?

_____ (1)

3 Give *one* reason why plants assimilate so little of the Sun's energy.

_____ (1)

4 Arrange the following to form a word equation:

  gross primary productivity   net primary productivity   respiratory energy loss

_____

_____ (1)

5 What is happening when something decays?

_____ (1)

6 In what form is energy lost as a result of respiration?

_____ (1)

7 In most ecosystems, what percentage of net primary productivity is converted into primary consumer productivity?

_____ (1)

8 Why is the conversion of primary consumer productivity into secondary consumer productivity more efficient?

_____ (1)

9 Why are food chains rarely more than four or five links long?

_____

_____ (1)

10 What does a pyramid of biomass illustrate?

_____

_____ (1)

(Total 10 marks)

CONCEPT TEST

1 The diagram shows the energy flow through a chicken.

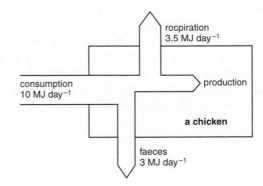

104

a Calculate the percentage of the energy in the chicken feed that ends up being absorbed across the chicken's gut.

(2)

b Battery chickens, which are kept in a small, centrally heated space, convert a greater percentage of the energy in chicken feed into primary consumer productivity than free-range chickens do. Why is this?

_____

_____

_____

_____
(3)

2 A pyramid of numbers shows the total number of organisms at each trophic level. A pyramid of biomass and a pyramid of numbers for the same food chain are shown below.

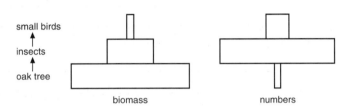

small birds
↑
insects
↑
oak tree

biomass   numbers

a Explain the shape of the pyramid of biomass.

_____

_____

_____
(3)

b Why does the pyramid of numbers have a different shape to the pyramid of biomass?

_____

_____

_____
(2)

(Total 10 marks)

NUTRIENT CYCLES

- The flow of energy through an ecosystem, which we looked at in the last unit, is a one-way process. Energy is absorbed as sunlight, is passed along the food chain, and is then lost, for good, as heat. The passage of vital chemicals, or **nutrients**, through an ecosystem is very different. When an organism takes up a nutrient, such as carbon or nitrogen, it does so only temporarily. Eventually, the nutrient will be released back into the environment, after which it can be taken up and used by another organism. The circular flow of a nutrient through an ecosystem, as it is repeatedly released and then reused, is known as a **nutrient cycle**.

- Living organisms are built out of organic molecules and all organic molecules are based on rings or chains of carbon atoms (see unit 1). This means that carbon is one of the most vital nutrients in the ecosystem. A summary of the **carbon cycle** is shown below.

- The main accessible source of inorganic carbon on this planet is the gas carbon dioxide (CO_2), which makes up 0.04% of the planetary atmosphere. In the world's oceans, CO_2 is available in dissolved form, as **hydrogencarbonate ions (HCO_3^-)**. Photosynthetic organisms take in, or **fix**, CO_2 and use it to construct organic molecules. The carbon in these organic molecules is then passed on to heterotrophs, such as animals and decomposers, when they eat the photosynthetic organisms. All organisms, both photosynthetic and heterotrophic, break down a proportion of the organic molecules that they have made, or eaten, in cellular respiration. As well as providing energy for the respiring organism, this breakdown results in the return of carbon to the environment, in the form of expired CO_2.

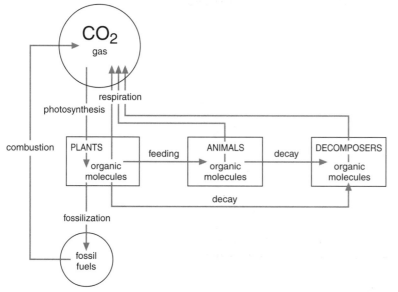

The carbon cycle

- In some environments, such as swamps, there is too little oxygen to support a population of decay bacteria. This means that when plants die, their organic molecules are not eaten by decomposers. Instead, they are gradually crushed and compacted, over millions of years, to produce carbon-rich **fossil fuels** such as coal and oil. Almost all of the coal and oil that has ever been formed has been burnt up by humans in the last two hundred years. The **combustion** of fossil fuels, on such a large scale, has resulted in a massive release of carbon dioxide and has led to global warming (see unit 28).

- Another vital nutrient is nitrogen, which is a necessary component of both proteins and nucleic acids (see units 3 and 10). An outline of the **nitrogen cycle** is shown left.

- Most plants take up nitrogen in the form of **nitrate ions (NO_3^-)**, dissolved in soil water. Having absorbed the nitrate ions through their roots, plants use them, and the nitrogen that they contain, to manufacture proteins. The **artificial fertilizers** used in modern farming are primarily composed of nitrate ions, since these help the crop plants to make more protein and thus to grow faster.

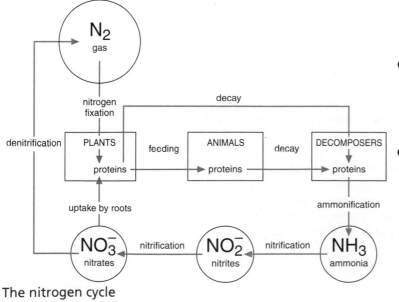

The nitrogen cycle

- Heterotrophs, such as animals and decomposers, obtain all the nitrogen that they need by eating the proteins contained in other organisms. These proteins are first broken down into amino acids, during digestion, and then the amino acids are absorbed (see unit 22). Often, after eating, a heterotroph will find itself in possession of more amino acids than it needs to build new proteins for its own body. One useful thing that it can do with the excess amino acids that it doesn't need is to break them down in respiration so that their energy can be used. Before amino acids can be respired, however, their nitrogen-containing component, the **amino group** (see unit 3), must be detached from them. This process is known as **deamination**. Once detached, the amino group becomes a molecule of **ammonia (NH_3)**. Humans convert this ammonia into **urea** and then excrete it in their urine. Most decomposers, such as bacteria, just excrete the ammonia directly. This means that when decomposers are feeding on protein, there will be a lot of ammonia released; a process known as **ammonification**.

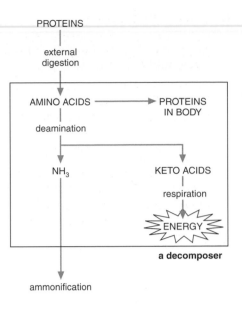

Ammonification in a decomposer

- There is a small amount of energy contained in the ammonia that is produced by ammonification. **Nitrifying bacteria**, in the soil, gain access to this energy by taking up the ammonia (NH_3) and oxidizing it to produce nitrites (NO_2^-) and nitrates (NO_3^-). Examples include *Nitrosomonas*, which converts ammonia into nitrites and *Nitrobacter*, which converts nitrites into nitrates. Nitrifying bacteria are autotrophs which, like plants, make their own organic molecules. Rather than using the energy in sunlight to do this, however, they use the energy that they get from the oxidation of the ammonia or the nitrites. For this reason, we call them **chemoautotrophs**. In a well aerated soil, where there is plenty of oxygen available to do oxidation with, nitrifying bacteria help to regenerate the nitrates which are needed by plants.

- Another important group of bacteria are the **denitrifying bacteria**. In a poorly aerated soil, where there is little oxygen available, denitrifying bacteria, such as *Pseudomonas*, take in nitrates (NO_3^-) and use the oxygen that they contain to drive respiration. They then give off nitrogen gas (N_2) as a waste product.

- Nitrogen gas (N_2) makes up 79% of the planetary atmosphere. Unfortunately, this huge supply of nitrogen is largely inaccessible to living organisms because the two atoms of nitrogen in a molecule of nitrogen gas are so tightly bonded together that they are almost impossible to break apart. This is why most plants have to rely on nitrates, in the soil, as a nitrogen source. There are a few species of bacteria, however, that possess an enzyme called **nitrogenase**. These **nitrogen-fixing bacteria** can use the nitrogenase that they contain to catalyse a reaction in which gaseous nitrogen is split apart and converted into ammonia. Some nitrogen fixers, such as *Azotobacter*, are free living, and these use the ammonia produced by fixation to help manufacture their own amino acids. Other nitrogen fixers, such as *Rhizobium*, live inside the **root nodules** of **leguminous plants**, such as sweet peas, beans, and clover. These nitrogen fixers provide their plant hosts with a continuous supply of ammonia, from which they can make proteins, in return for food, in the form of glucose, and housing. This is an example of a mutualistic relationship, in which two different organisms help each other out.

RECALL TEST

1 How does the movement of nutrients through an ecosystem differ from the movement of energy?

_____

_____

(1)

2 In what form do marine plants take up carbon?

_____ (1)

3 What process is involved in carbon fixation?

_____ (1)

4 Name a process, other than respiration, that results in the release of CO_2.

_____ (1)

5 What do artificial fertilizers contain?

_____ (1)

6 What happens during deamination?

_____

_____

(1)

7 Name a species of nitrifying bacteria.

_____ (1)

8 Why are nitrifying bacteria described as 'chemoautotrophs'?

_____

_____

(1)

9 What do denitrifying bacteria release as a waste product?

_____ (1)

10 What reaction is catalysed by the enzyme nitrogenase?

_____

_____

(1)

(Total 10 marks)

CONCEPT TEST

1 Insect-eating plants, such as sundew and the Venus fly trap, use sticky or jaw-like leaves to catch and digest insects that alight on them. These plants tend to be found growing in swampy or waterlogged regions.

    a Insect-eating plants can make all the carbohydrates they need by photosynthesis. How, then, do they benefit from eating insects?

_____

_____

(2)

b Why is insect eating of particular benefit to a plant growing in a waterlogged soil?

_____

_____

_____

_____

_____ (3)

2 The graphs below show the relative abundance of leguminous and non-leguminous plants found growing on two types of soil.

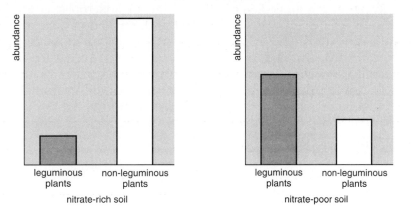

a Why do the leguminous plants prosper better than the non-leguminous plants on the nitrate-poor soil?

_____

_____

_____

_____ (2)

b Why are leguminous plants less abundant on the nitrate-rich soil than they are on the nitrate-poor soil?

_____

_____

_____

_____

_____ (3)

(Total 10 marks)

HUMAN IMPACT ON THE ENVIRONMENT

● All living organisms have an effect, or impact, on the environment. Plants absorb carbon dioxide from the atmosphere and pump out the water that they've drawn up from the ground. Animals excrete and compete and eventually decay, adding their nutrients to the soil. From the perspective of the human life span, all these various impacts may appear to balance out, creating a miraculously stable planetary ecosystem, but this is, in fact, an illusion. Viewed over the long term, in millennia rather than decades, the story of life on this planet is a tale of flux and instability, of asteroid impact and sudden climate change, of mass extinction, collapse, and then recovery.

● Four billion years ago, the Earth was inhabited by a complex community of anaerobic creatures. In those days, there was no oxygen in the atmosphere at all. When the first photosynthetic organisms evolved and began to pump out oxygen, it was an ecological disaster for the anaerobes. From their point of view, oxygen was a poison, and most of them died out. Sixty million years ago, the dinosaurs, who had ruled the earth for over a hundred million years, vanished entirely from the face of the planet. Were they wiped out by a meteorite impact, or by volcanic activity, or by a rapidly changing climate? We still don't really know. Three million years ago, all the pouched, or marsupial, mammals in South America were driven to extinction by an invasion of non-pouched mammals. Only a few hundred thousand years ago there were crocodiles and hippos living in the river Thames. Then an ice age came along.

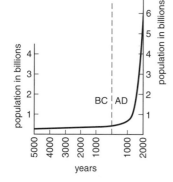

Human population growth

● As you can see, the Earth has certainly experienced instability and disruption before. In the last couple of thousand years, however, it has been suffering particularly badly at the hands of a new and uniquely disruptive species: the human being. Humans use intelligence and technology to alter the environment. This manipulative ability, as exemplified by the invention of agriculture and industry, has allowed the human population to increase exponentially. The bigger the human population gets, the more it demands from the environment and the bigger its impact.

● The first humans lived in small, wandering bands, as hunter-gatherers. On the whole planet, there were fewer people than you'd find today in a single city. Even at this low population level, mankind had an effect on the environment. There is evidence that the arrival of the first human hunters into North America was followed by the extinction of many mammal species. Early Australian hunters even set fire to large areas of forest, in order to produce the kind of grassland environment which would encourage the breeding of big game animals.

● It was with the origin of cities and agriculture, however, that human impact really began to become extreme. In order for crops to be planted, the vegetation that occurs naturally in an area has to be cleared. In wooded regions this leads to **deforestation**. As well as resulting in the loss of many different tree species, deforestation also eradicates animal species that depend on the trees for food and for a place to live. In addition to this loss of **biodiversity**, there may also be a loss of soil quality. The roots of trees help to hold soil together and prevent rain from beating down on it. When **rainforest** is cleared, usually for lumber, much of the soil that's left is soon washed away or has all the nutrients drained out of it. This process, known as **leaching**, eventually renders the ground unfit for even the most basic agriculture.

● Once crops have been planted, there are a number of farming practices, used to help the crops grow well, which can further impact on the environment. **Irrigation**, the diversion of water from its natural source to where the crops are growing, can leave lakes stagnant and reduce rivers to a trickle. The spraying of **herbicides** and **pesticides**, to kill competing weeds or insects that might feed on the crops, can lead to the poisoning of other plants and animals. The addition of artificial fertilizers (see unit 27), to improve crop yield, can lead to the eutrophication of neighbouring rivers and lakes.

- **Eutrophication** begins when the run-off of water from a field carries fertilizer into a nearby body of water. Here, the fertilizer causes the rapid growth of aquatic plants, such as algae, leading to an **algal bloom**. As dead algae accumulate on the bed of the river or lake, the number of decay bacteria increases dramatically. With a huge amount of bacterial respiration going on, all the oxygen in the water is soon used up. Fish, which need oxygen themselves, can rarely survive this kind of impact.

- Since the industrial revolution, in the nineteenth century, the impact of factory manufacture on a global scale has been added to that of agriculture. Manufacture of consumer goods such as cars, televisions, etc. depends on the burning of fossil fuels (see unit 27). Fossil fuels must also be burned in order to provide the energy that is needed to drive the cars and power the televisions. There are a number of negative consequences that result from large-scale use of fossil fuels. **Oil slicks**, released by damaged oil tankers, can destroy miles of coastline. **Acid rain**, formed when the products of fossil-fuel combustion mix with rainwater, can destroy whole forests. **Dust particles** and **carbon monoxide**, from car exhausts, can make the air in a city dangerous to breathe. The most significant impact of fossil-fuel use, however, is global warming.

- **Global warming** results from the industrial emission of **greenhouse gases**, such as carbon dioxide, chlorofluorocarbons, methane, and ozone. These gases, once released into the atmosphere, act like the panes of glass in a greenhouse. They allow light in, through the atmosphere, but prevent heat from escaping. Even if all fossil-fuel consumption stopped now, the greenhouse gases that have already been released would continue to contribute to a rise in planetary temperature. With increasing fossil fuel use, which is what seems likely, the global average temperature is set to rise by up to 6° in the next hundred years. The two major effects that this will trigger are a rise in global sea level, caused by melting of the Antarctic ice cap, and a disruption of global weather patterns. Expect more floods, droughts, tornadoes, etc.

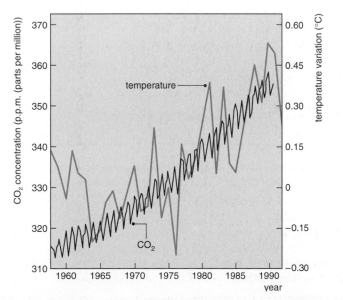

The increase in atmospheric CO$_2$ and temperature from 1958 to 1991

- Chlorofluorocarbons, as well as being greenhouse gases, have the additional effect of destroying the **ozone layer**, which protects the Earth from harmful **ultraviolet radiation**. Other **pollutants** released by industry include **heavy metals**, **radioactive** substances, and **carcinogenic**, or cancer-causing, substances. The list is endless. As a final point, the recent development of **genetic engineering** (see unit 12) offers humans a yet more powerful way to disrupt the environment, through the release of genetically modified organisms.

TESTS

RECALL TEST

1 Why does deforestation lead to a loss of animal biodiversity?

_____

_____ (1)

2 What is 'leaching'?

_____

_____ (1)

3 What agricultural practice results in eutrophication?

_____ (1)

4 Name two other agricultural practices that can have a negative impact on the environment.

_____ (1)

5 How is acid rain formed?

_____

_____ (1)

6 Why is the air in a city dangerous to breathe?

_____

_____ (1)

7 Name three greenhouse gases.

_____

_____ (1)

8 What are the harmful effects of global warming?

_____

_____ (1)

9 Which gases are responsible for destroying the ozone layer?

_____ (1)

10 Why is the loss of the ozone layer a problem?

_____

_____ (1)

(Total 10 marks)

CONCEPT TEST

1 The table below shows figures relating to the recycling of plant biomass in two different ecosystems.

	living plant biomass ($kg\,m^{-2}$)	new plant material per year ($kg\,m^{-2}$)	organic matter in soil ($kg\,m^{-2}$)
Deciduous woodland	40.7	0.9	1.5
Tropical rainforest	52.5	3.3	0.2

a Why is the production of new plant material highest in the tropical rainforest?

_____

_____

_____

_____

           (2)

b Given that a greater quantity of plant litter is produced each year in the rainforest, why does its soil contain the least organic matter?

_____

_____

_____

           (2)

c Use the figures in the table to explain why the deforestation of rainforest has more of a long-term impact than the deforestation of deciduous woodland.

_____

_____

_____

_____

           (2)

2 The graph below shows some of the changes that occur within a river, downstream from a sewage outlet.

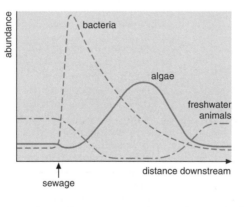

a Why are there more algae downstream from the outlet?

_____

_____

           (2)

b Why are there fewer freshwater animals downstream from the outlet?

_____

_____

_____

           (2)

           (Total 10 marks)

ANSWERS

Throughout this section, model answers are given in regular text. Hints, advice, and extended explanations are written in italics.

UNIT 1

RECALL TEST

1 A molecule is a collection of atoms which are joined together by bonds.
2 An element is a substance that contains only one type of atom.
3 4.
4 N_2, O_2, CO_2, and H_2O.
5 Two hydrogen and one oxygen, giving H_2O.
6 A glycosidic bond.
7 Amino acids.
8 Oxygen and nitrogen.
9 A covalent bond.
10 Cellulose and DNA.

CONCEPT TEST

1 N_2H_4.
2 3. Because, in the molecules shown, a nitrogen atom always makes 3 covalent bonds with its neighbours.
3 The structural formulae do not provide any information about the three-dimensional shape of the molecules.
4 A and C. Because their structures are based on chains of carbon atoms.
5 A is water soluble and C isn't. This is because A has OH and NH groups, which will give it charge and allow it to form hydrogen bonds with water, while C doesn't.

UNIT 2

RECALL TEST

1 $C_3H_6O_3$.
2

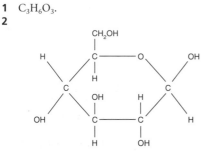

3 Glucose and fructose.
4 *Any three from:* sweet, white, crystalline, soluble in water.
5 It is insoluble and compact.
6 Cellulose.
7 It consists of a glycerol head and three fatty-acid tails.
8 They lack OH or NH groups, which means that they cannot hydrogen-bond with water.
9 Unsaturated fatty acids have some double bonds between their carbon atoms, saturated fatty acids don't.
10 *Any five from:* energy storage, insulation, protection of delicate organs, component of cell membranes, waxy coating of leaf, sex hormones.

CONCEPT TEST

1 Molecule A: glycogen; molecule B: cellulose; molecule C: amylose; molecule D: sucrose; molecule E: deoxyribose.
2 *Maltose is made of two glucose molecules joined together by a condensation reaction in which water is removed. So:*
$C_6H_{12}O_6 + C_6H_{12}O_6 - H_2O = C_{12}H_{22}O_{11}$
3 Because carbohydrates are easier to break down, they are the ideal energy-storage molecule. Since animals are mobile, however, and can't afford to be too heavy, they have to store some energy in the form of lipids, because these contain more energy per gram of weight. Plants, which aren't mobile and don't have to worry about weight, just store carbohydrates.

UNIT 3

RECALL TEST

1 Nitrogen and sulphur.
2

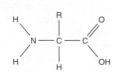

3 20.
4 The R group.
5 A condensation reaction.
6 A peptide bond.
7 The order in which the amino acids are joined together to make the polypeptide chain.
8 α-helix and β-pleated sheet.
9 Hydrogen bonds, disulphide bonds, electrovalent bonds.
10 Haemoglobin.

CONCEPT TEST

1 a A polypeptide chain. / A chain of amino acids.
   b An amino group (NH_2).
   c Quaternary structure. Because the protein is composed of several polypeptide chains stuck together.
2 *There are 20 choices for each amino acid in the chain of five, so:*
$20 \times 20 \times 20 \times 20 \times 20 = 3\,200\,000$
3 Every type of receptor molecule has to have a different shape, because each type has to fit onto and recognize a different sort of messenger molecule. Proteins are the ideal molecules to use as receptors, because, with so many different ways of putting amino acids together, they can be designed to have almost any shape that's needed.

UNIT 4

RECALL TEST

1 Protein.
2 Because they increase the rate of chemical reactions.
3 The energy that has to be put in to start a reaction off.
4 enzyme + substrate → enzyme–substrate complex → enzyme + end product
5 Because each enzyme has an active site that will only fit onto a particular substrate.
6 A process in which the active site changes shape once the substrate has fitted into it.
7 Hydrogen bonds.
8 High temperature. Extreme pH.
9 Cyanide.
10 They block the active site of the enzyme and so prevent the substrate from fitting in.

CONCEPT TEST

1 a The reaction at 60 °C is running faster than that at 40 °C because, at a higher temperature, there are more collisions between enzymes and substrate molecules. As a result, the end product accumulates quicker at 60 °C than at 40 °C.
   b After 6 minutes have passed, the enzymes at 60 °C have denatured, due to the high temperature, and end product is no longer being produced. The enzymes at 40 °C, meanwhile, are continuing to catalyse the reaction and end product is continuing to accumulate.
2 a Initially, as the substrate concentration is increased, the reaction rate increases. Eventually, however, the enzymes are all working as fast as possible and adding more substrate has no effect, which is why the graph levels off.

b The addition of a non-competitive inhibitor slows the reaction rate at all substrate concentrations. A competitive inhibitor, however, has little effect at high substrate concentrations. This is because at high substrate concentrations the competitive inhibitor is out-competed by the substrate in its competition for the enzyme's active site. *i.e. if there are lots of substrate molecules around, a competitive inhibitor rarely gets a chance to get into an active site before a substrate molecule does.*

UNIT 5

RECALL TEST
1 Eukaryotic cells contain membrane-bound organelles. Prokaryotic cells don't.
2 An infolding of the prokaryotic cell membrane, to which vital enzymes are attached.
3 A small loop of DNA found in a prokaryote.
4 Plasmodesmata.
5 Nucleus. Mitochondrion. Chloroplast.
6 Inside the nucleus.
7 The rough endoplasmic reticulum has ribosomes attached to it. The smooth endoplasmic reticulum doesn't.
8 It modifies proteins and other molecules and then packages them into transport vesicles.
9 Digestive enzymes.
10 A biconcave disc.

CONCEPT TEST
1 *A real animal cell is 20 μm across. If we imagine this expanded to 5 m across (which is 5000 mm or 5000 000 μm), that is a magnification of 5000 000/20, or 250 000. In order to work out the sizes of the various organelles in an imaginary 5 m cell, therefore, all we have to do is multiply their actual sizes by 250 000.*

Size of nucleus	$5\,\mu m \times 250\,000$	$= 1250\,000\,\mu m = 1.25\,m$
Size of a mitochondrion	$1\,\mu m \times 250\,000$	$= 250\,000\,\mu m = 250\,mm$
Size of a ribosome	$15\,nm \times 250\,000$	$= 3750\,000\,nm = 3.75\,mm$
Width of a cell membrane	$8\,nm \times 250\,000$	$= 2000\,000\,nm = 2\,mm$

This means that, in an animal cell as big as a room, the cell membrane would only be 2 mm thick, about the width of a thin piece of cardboard, while the nucleus would be the size of a sofa. It is always worth trying to visualize the real dimensions of biological structures.

2

	Contains chloroplasts	Has a cell wall	Contains ribosomes	Contains centrioles	Has a nucleus
Bacterial cell	✗	✔	✔	✗	✗
Plant cell	✔	✔	✔	✗	✔
Animal cell	✗	✗	✔	✔	✔

3 Prokaryotic cells, like mitochondria, have a loop of DNA and 70S ribosomes. This suggests that mitochondria may once have been independent prokaryotic organisms, rather like bacteria, that ended up being engulfed by, and becoming part of, another cell. *This theory is known as the 'endosymbiotic theory' of eukaryote origin.*

UNIT 6

RECALL TEST
1 The fact that they have polar heads and non-polar tails.
2 Because both proteins and phospholipids can float around freely within the plane of the membrane.
3 Enzymes. Receptors. Molecular pumps.
4 A protein with a bit of carbohydrate attached to it.
5 When equilibrium has been reached.

6 They are charged and therefore cannot enter the uncharged region of the membrane where the fatty-acid tails are.
7 An uncharged molecule such as a fat.
8 Because they will only transport molecules that fit into their binding site.
9 Because it requires no energy expenditure by the cell.
10 In order to transport molecules against a concentration gradient so that they can be accumulated on one side of a membrane.

CONCEPT TEST
1 The dark bands are produced by rows of phospholipid heads. The light region in the middle is where the phospholipid tails are.
2 **a** From the left, where the total concentration of sugars is $0.36\,mol\,dm^{-3}$, to the right, where it is $1.11\,mol\,dm^{-3}$. *In osmosis, water moves from where there are fewer dissolved substances to where there are more.*

  b Glucose will move from right to left. Fructose will move from left to right. *Maltose will not be able to cross the membrane since it is a disaccharide. Both monosaccharides will diffuse from where they are in higher concentration to where they are in lower concentration.*

3 Since poisoning the cell has no immediate effect, this must be a passive transport process that does not require energy production, i.e. it is not active transport. When external glucose concentrations get very high, the uptake rate stops increasing, which implies that we are looking at facilitated diffusion, in which the number of carrier proteins acts as a limiting factor. *In facilitated diffusion the transport proteins can only work so fast. Eventually, adding extra substrate will have no effect, since all the transport proteins will be working as fast as possible.*

UNIT 7

RECALL TEST
1 From blue to brownish red.
2 Iodine.
3 Potassium hydroxide. Copper sulphate.
4 R_f value of molecule $= \dfrac{\text{distance moved by molecule}}{\text{distance moved by solvent}}$
5 In order to break down the cellulose cell walls.
6 The same concentration.
7 The material that doesn't sediment out to form a pellet after being spun in centrifugation.
8 nuclei + mitochondria, ribosomes + membrane fragments, soluble material.
9 Resolution is the ability to distinguish detail. Magnification is the ability to make something appear larger.
10 Because the electrons in an electron beam are smaller than the photons in a beam of light.

CONCEPT TEST
1 **a** Substance **A** must be sucrose, since boiling it with HCl and then performing a Benedict's test is the only procedure that produces a positive result.

  b Substance **B** must be an enzyme that breaks sucrose down, since, in combination with sucrose, it results in the production of reducing sugars. *We know this because the combination of **A** and **B** gives a positive result when a simple Benedict's test is performed. The fact that **B** gives a positive result with a Biuret test confirms that, like all enzymes, it is a protein.*

  c Since substance **A** is sucrose, it is clear that it must be sucrose in position **X**. The molecules found in positions **Y** and **Z** must be the glucose and fructose that the sucrose is broken down into when the enzyme is added.

2 **a** A scanning electron microscope. *As mentioned in the text, this gives a three-dimensional surface view, which would be ideal for viewing the surface of a pollen grain. A light microscope could also be used but would give a much less detailed view.*

   b A light microscope. *Electron microscopes cannot be used to observe things while they are still alive.*

   c A transmission electron microscope. *This would give the kind of cross-sectional view needed to see inside a chloroplast as well as providing enough resolution to make out the details.*

UNIT 8

RECALL TEST

1 Inside molecules of ATP.
2 It generates more ATP. It can be used to get energy out of many different types of molecule, not just carbohydrates. It doesn't produce a poisonous by-product.
3 In the cytoplasm of the cell.
4 The addition of phosphates, which destabilizes the molecule.
5 It allows the regeneration of NAD from NADH + H$^+$.
6 Acetyl CoA.
7 On the inner membrane of the mitochondrion.
8 From glycolysis. From the conversion of pyruvate into acetyl CoA. From the Krebs cycle.
9 Oxygen.
10 36 ATP.

CONCEPT TEST

1 The total amount of alcohol increases, initially, as it is produced by the growing yeast population. Since alcohol is a poison, however, it eventually kills off the yeast population and no more is produced, which is why the graph levels off.
2 The anaerobic breakdown of a glucose molecule produces far less ATP than its aerobic breakdown. This means that, in order to supply the energy demand of the ripening process, much more glucose has to be broken down in anaerobic conditions and this produces more CO_2.
3 **a** Inside the mitochondria of a respiring yeast cell, hydrogens are pulled off a respiratory substrate, as it is broken down in the Krebs cycle. Normally these hydrogens are passed to the electron transport chain, so that ATP can be made, but in this case they are passed to the methylene blue, causing it to change colour.

   b Shaking the test tube helps to mix in oxygen. This oxygen oxidizes the methylene blue, pulling hydrogens off it and causing it to change colour.

UNIT 9

RECALL TEST

1 Plants. Single-celled algae. Cyanobacteria.
2 Because they are 'self-feeders' that manufacture their own organic molecules.
3 From water.
4 In the grana of the chloroplasts.
5 ATP and NADPH + H$^+$.
6 Ribulose biphosphate.
7 5.
8 The splitting of water.
9 The process in which ATP is produced during the light-dependent stage of photosynthesis.
10 In order to boost an electron to an energy level high enough for it to be picked up by NADP.

CONCEPT TEST

1 By trapping the reactants involved in photosynthesis within the small space of a chloroplast, all of them can be kept at a high concentration, which means that they are more likely to collide, interact, and react with one another.

2 **a** Water, which is split in photolysis.

   b They are passed to a photosystem, or collection of chlorophyll molecules, where light energy is absorbed and used to 'boost' the electrons to a higher energy level.

   c They would be passed from photosystem II to photosystem I and then on to NADP which would carry them into the Calvin cycle, where they would end up becoming part of a glucose molecule.

3 When the chloroplast is moved into the dark, ATP and NADPH + H$^+$ can no longer be provided by the light-dependent stage and therefore glycerate-3-phosphate cannot be converted into triose phosphate. With the Calvin cycle interrupted in this way, glycerate-3-phosphate will accumulate and the triose phosphate reserve will decrease. *If you look at a Calvin cycle diagram you can see that the triose phosphate will get turned into RuBP, which will then get turned into glycerate-3-phosphate. Since the glycerate-3-phosphate can't be turned into triose phosphate without ATP and NADPH + H$^+$, it will accumulate and the triose phosphate, which isn't being replaced, will decrease.*

UNIT 10

RECALL TEST

1 Deoxyribonucleic acid.
2 A phosphate group, a pentose sugar, and a nitrogenous base.
3 Adenine and guanine.
4 Phosphodiester bonds.
5 Adenine–thymine. Guanine–cytosine.
6 Because the two chains lie side by side and run in opposite directions.
7 A piece of DNA and the histone proteins that it is wound around.
8 DNA contains deoxyribose sugar, as opposed to ribose in RNA. DNA contains the base thymine, as opposed to uracil in RNA. DNA is composed of two polynucleotide chains, as opposed to one in RNA.
9 DNA polymerase.
10 The complementary nature of base pairing during replication.

CONCEPT TEST

1 **a** If, in a DNA double helix, there is a purine base on one strand, this will always be connected to a complementary pyrimidine base on the other strand. This means that number of purine bases (A + G) will always be roughly equal to the number of pyrimidines (T + C).

   b In mRNA there is only one strand, with no pairing between bases, and therefore there is no reason why the proportion of purines should equal that of pyrimidines.

   c The fact that the squirrel and the shark have similar proportions of the four bases is irrelevant. It is the different **sequence** in which the bases occur in a squirrel and in a shark that makes them different.

2 According to the semiconservative theory of DNA replication, each new piece of DNA double helix is constructed from one old polynucleotide strand and one new strand. This view is supported by the experiment, in which the new pieces of DNA are of intermediate weight, indicating they are made of old, heavy, ^{15}N strands combined with new, light, non-labelled strands.

UNIT 11

RECALL TEST

1 DNA → transcription → translation → enzyme → chemical reaction → phenotype
   The phenotype of an organism is its physical makeup, i.e. what sort of creature it is.

2   A small section of a chromosome that carries information about one particular polypeptide.

3   A group of three bases that code for one particular amino acid.

4   64.

5   Because the same codons are used, in all living organisms, to code for the same amino acids.

6   RNA polymerase.

7   Hydrogen bonds. Phosphoester bonds.

8   In the rough endoplasmic reticulum.

9   Protein and rRNA.

10 Molecules of tRNA are used to carry amino acids into place during translation.

CONCEPT TEST

1   a   AUA.
        The DNA codon is ATA. This means that the codon on the mRNA copied from the DNA will read UAU. The complementary anti-codon on the tRNA will thus be AUA.

    b   Because the genetic code is degenerate, all the codons beginning GC code for the amino acid arginine and therefore changing the base found in position **X** will have no effect on the polypeptide produced.

    c   Deleting the base found in position **X** will cause a 'frame shift'. The first codon will now be read as GCC, and the next as CCA, etc. Since every subsequent codon will be altered in this way, including the stop codon, the gene will not terminate after 4 codons but will carry on until, after 15 codons, a new and randomly created stop codon is encountered.

2   The sense and nonsense strands of a section of double-helical DNA are complementary. This means that the pieces of mRNA copied from the sense strand and from the nonsense strand will also be complementary and will therefore bind together. With no exposed mRNA bases available, tRNA molecules will have nothing to attach to during translation and no polypeptide will be produced.

UNIT 12

RECALL TEST

1   DNA which has had new bits added in.

2   The use of genetically engineered bacteria to produce insulin *or* the use of genetically modified yeast to produce chymosin.

3   The treatment of human hereditary diseases using genetic engineering.

4   In order to protect themselves against invading viral DNA.

5   Because each type has a different active site that recognizes and fits onto a different sequence of DNA bases.

6   DNA that has been copied from mRNA by reverse transcriptase.

7   Something that helps to carry a gene into a target organism.

8   Because they naturally swap plasmids between each other and are therefore used to picking plasmids up.

9   It helps to join two pieces of DNA together, end to end.

10 Microinjection.

CONCEPT TEST

1   a   The arrow should point from **right** to **left**. *Smaller fragments always move furthest during electrophoresis and since the smallest fragments (2 kilobases) are on the left, this tells us that the current must have been flowing from right to left.*

    b   Each band consists of a group of restriction fragments, or pieces of DNA, of identical length.

    c   A different restriction enzyme would have a different active site and would therefore fit onto and cut the DNA at different points, producing a set of restriction fragments with different lengths. This would give rise to a different banding pattern.

2   a   A staggered cut leaves a strand of unpaired bases projecting out from each cut end. These unpaired bases are complementary to one another and will have a tendency to recombine, making the cut ends 'sticky'.

    b   Pieces of DNA that are cut by restriction enzymes that leave sticky ends will be able to join easily to other pieces of DNA with complementary sticky ends. This is useful when genetic engineers want two pieces of DNA to join together, e.g. when they want a cut open plasmid to rejoin so that it incorporates a new gene.

UNIT 13

RECALL TEST

1   The division of a cell to produce two identical daughter cells.

2   Growth. Repair. Asexual reproduction.

3   Interphase, prophase, metaphase, anaphase, telophase.

4   Metaphase.

5   Prophase.

6   Prophase.

7   An organelle that controls the formation of spindle fibres.

8   The structure that holds two chromatids together.

9   Because the chromosomes uncoil or uncondense to become a tangled mass of chromatin again.

10 The process in which a cell splits in two, by pinching in at the sides.

CONCEPT TEST

1   Because anaphase is the shortest of the five phases.

2   Because the whole phase looks the same through a light microscope. All that can be seen is a tangled mass of chromatin.

3   Because a cell is roughly spherical in shape, like the globe of the Earth, which we describe as having 'poles' and an 'equator'.

4   a   Anaphase is beginning. *It is at this point that the distance between the chromatids begins to increase, which must mean that they are being pulled apart, i.e. anaphase is beginning.*

    b   The distance between the chromatids and the poles of the cell. *This distance begins to **decrease** at the start of anaphase, so it must be the distance between chromatids and poles.*

5   If colchicine is added, all the cells will get to prophase, with their chromosomes condensed and neatly visible, and will then stop there, since no spindle fibres are available to initiate metaphase or anaphase.

UNIT 14

RECALL TEST

1   The division of a diploid parent cell to produce four, non-identical, haploid daughter cells.

2   Because the daughter cells produced by meiosis only contain half the normal amount of DNA.

3   The production of gametes.

4   Because they come from different parents.

5   *Any three from:* the chromosomes condense and become visible. The nuclear membrane breaks down. Centrioles move to the poles of the cell. Spindle fibres form. The chromosomes begin to be pulled towards the cell equator.

6   In anaphase I chromosomes line up in a single row down the equator. In anaphase II they line up in homologous pairs.

7   Because it includes two complete cell divisions.

8   AB. A'B. AB'. A'B'.

9   During prophase and metaphase of meiosis I.

10 It creates chromosomes that contain new combinations of genes.

CONCEPT TEST

1
a Mitosis. *You can tell that this is mitosis because the DNA content doubles and is then halved back to normal.*

b Meiosis. *You can tell that this is meiosis because the DNA content doubles and is then halved twice in a row, to produce cells which are haploid, with half the normal DNA content.*

c The division, or cytokinesis, of the cell after anaphase of meiosis I.

d Gametes. *Since they are produced by meiosis and they have only half the normal DNA content.*

e Fertilization, or fusion of gametes. *Once gametes have been produced, the only way to get back to the normal, diploid quantity of DNA is for two gametes to fuse together.*

2

	Centromeres are split	A spindle is formed	Homologous chromosomes pair up together	DNA replication occurs at the start of the division	Crossing over occurs
Mitosis	✔	✔	✗	✔	✗
Meiosis I	✗	✔	✔	✔	✔
Meiosis II	✔	✔	✗	✗	✗

3 6.
If there are 6 chromosomes in anaphase II,

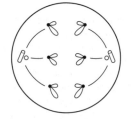

there must have been three in metaphase II, before the chromatids were pulled apart.

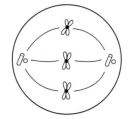

This means that there must have been three pairs lined up down the middle in metaphase I, before these pairs were pulled apart. This gives us 6 chromosomes.

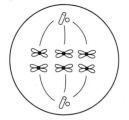

UNIT 15

RECALL TEST

1 The genetic makeup of an organism/the genes that an organism inherits from its parents.
2 The physical characteristics of an organism.
3 The position at which a gene occurs on a chromosome.
4 Different versions of a gene that can occur at a particular locus.
5 ABO blood group.
6 That it has inherited two different alleles for that characteristic.
7 'Short tail' is dominant. 'Long tail' is recessive.
8 Codominance.
9 1. *Because gametes, such as a sperm cell, are haploid and contain only one allele for each characteristic.*

10 Because the ratios predicted by Mendelian genetics are only statements of probability. Actual offspring ratios are subject to chance.

CONCEPT TEST

1
a 2.
b $c^{ch}c^{ch}$, $c^{ch}c^{h}$, $c^{ch}c$.
c

Phenotype of parents	albino rabbit × himalayan rabbit	
Genotype of parents	cc	$c^{h}c$
Genotype of gametes	(c)	(c^{h}) or (c)

Genotype and phenotype of offspring		(c^{h})	(c)	in a 1:1 ratio of himalayan: albino
	(c)	$c^{h}c$ himalayan	cc albino	

2
a Blood group B. *Individual 4 has blood group B. She must, therefore, have inherited an I^{B} allele from one of her parents. She can't have inherited this allele from her mother, who is blood group O ($I^{O}I^{O}$), therefore she must have inherited it from her father. This tells us that individual 2 has one I^{B} allele. Individual 3 is blood group O and must, therefore, have inherited an I^{O} allele from each of his parents. This tells us that individual 2 has one I^{O} allele. Since we have deduced that individual 2 has one I^{B} and one I^{O} allele, we know that they have blood group B.*

b $I^{A}I^{O}$. *Individual 5 must have at least one I^{A} allele, in order to be blood group A. The second allele could either be another I^{A} or an I^{O}. It cannot be an I^{A} because this would result in offspring with blood group AB. Therefore, individual 5 must have the genotype $I^{A}I^{O}$.*

c $\frac{1}{8}$. *We know that individual 4 has the genotype $I^{B}I^{O}$, since she is blood group B and can only have received an I^{O} from her mother. We know that individual 5 has the genotype $I^{A}I^{O}$. If we draw a Punnett square to work out the expected ratio of offspring produced when individuals 4 and 5 mate,*

	(I^{A})	(I^{O})
(I^{B})	$I^{A}I^{B}$	$I^{B}I^{O}$
(I^{O})	$I^{A}I^{O}$	$I^{O}I^{O}$

we can see that there is a $\frac{1}{4}$ chance of the next child having blood group B. The chance of the next child being a girl is, of course, $\frac{1}{2}$. Combining these probabilities together, we can calculate that the chance of the next child having blood group B and being a girl is $(\frac{1}{4}) \times (\frac{1}{2})$, which is equal to $\frac{1}{8}$.

UNIT 16

RECALL TEST

1 Single-celled prokaryotes.
2 A group of organisms that can breed together to produce fertile offspring but cannot breed with other organisms.
3 The evolutionary history of, and relationships between, a number of different species.
4 The gradual change in the nature of organisms within a species, over time.
5 Due to mutation and to the random shuffling of genes that occurs during meiosis and sexual reproduction.
6 We mean that organisms which are better suited to their environment are the ones that survive to pass on their genes.
7 A directional change in the average value of a characteristic.
8 When a species cannot change fast enough to keep up with a rapidly changing environment.
9 A process in which an old species splits in two to produce two groups of organisms that cannot interbreed.
10 Geographical isolation.

CONCEPT TEST

1 a The seeds planted on normal soil show a large range of different copper tolerances due to mutation and to the random shuffling of genes that occurs due to meiosis and sexual reproduction.

  b The plants growing on the copper mine show a narrower range of tolerance because only those seeds with a high tolerance have survived and grown to become adult plants.

  c Natural selection/directional selection.

2 a They must all have evolved from a common ancestor species.

  b Each species has been isolated in a different area, leading them to evolve slightly different physical characteristics.

  c The advance of the ice sheet must have split the original Warbler population into three isolated groups, leading to allopatric speciation. Though the ice sheet has since retreated, it has left behind it the diversity that we see today.

UNIT 17

RECALL TEST

1 A group of organisms that can breed together to produce fertile offspring but cannot breed with other organisms.

2 Because it involves giving every species a name that has two parts.

3 *Homo sapiens*.

4 The science of dividing species up into groups, or classifying them.

5 Kingdom Prokaryotae.

6 An organism that lives inside, and feeds off, another organism.

7 *Amoeba* or *Paramecium*.

8 A flowering plant.

9 Kingdom Protoctista.

10 Yeast.

CONCEPT TEST

1 The official definition of a species is a group of organisms that can breed together but that cannot breed with other groups. The problem when classifying fossil organisms is that there is no way of working out which other organisms they could breed with, since they are dead.

2

Kingdom	Phylum	Class	Order	Family	Genus	Species Name
Animalia	Chordata	Mammalia	Carnivora	Felidae	*Felis*	*Felis silvestris*

3 It is the degree of relatedness between two organisms that taxonomists pay attention to when they are classifying them. Since DNA differences are indicator of relatedness, with closely related organisms having fewer DNA differences, they are a useful tool for taxonomists.

4

	Contains single-celled organisms	Contains organisms built out of hyphae	Contains photosynthetic organisms	Contains eukaryotic organisms	Contains organisms with cellulose cell walls
Kingdom Prokaryotae	✔	✘	✔	✘	✘
Kingdom Protoctista	✔	✘	✔	✔	✔
Kingdom Plantae	✘	✘	✔	✔	✔
Kingdom Animalia	✘	✘	✘	✔	✘
Kingdom Fungi	✔	✔	✘	✔	✘

UNIT 18

RECALL TEST

1 Aerobic respiration.

2 By simple diffusion across its cell membrane.

3 It decreases.

4 The branching tubes that allow gas exchange in insects.

5 Stomata.

6 $\text{rate of diffusion} = \dfrac{\text{surface area} \times \text{difference in concentration}}{\text{thickness of membrane}}$

7 The pair of tubes that the trachea divides into, in the lungs.

8 Because they contain over 700 million alveoli.

9 In order to maintain a steep oxygen concentration gradient between the alveoli and the blood.

10 It decreases.

CONCEPT TEST

1 Because the lugworm is a particularly large variety of worm it has a particularly small surface area-to-volume ratio. This means that, unlike smaller worms, it cannot get all the O_2 it needs by simple diffusion across its body surface. The external gills compensate for this by providing extra surface area for oxygen uptake.

2 In terms of Fick's law, a gas exchange surface maximizes gas exchange by having a large surface area, by maintaining a steep concentration gradient, and by being thin. The many gill filaments in a fish's gill create a large surface area for gas exchange. The continuous movement of water across the gill filaments helps to maintain a steep concentration gradient, by continuously replacing old, stale water with new, oxygenated water. Finally, because the gill filaments are filled with blood, there is only a thin gas exchange layer separating the water from the blood.

3 A person can absorb oxygen from the first lungful of water that they breathe in but, because human lungs are not designed to inhale and exhale water, they cannot take a second breath. This means that they cannot replace old, stale water with new, oxygenated water and thus they drown.

UNIT 19

RECALL TEST

1 The right ventricle.

2 Between the left atrium and the left ventricle.

3 They prevent the flaps of the atrioventricular valves from opening backwards.

4 The impact of blood against the atrioventricular valves when they first close.

5 Both the atrium and ventricle relax and gradually fill with blood.

6 Because it has to pump blood all the way around the body, while the right ventricle only has to pump blood to the lungs.

7 In the wall of the right atrium.

8 It allows electrical activity to pass through the atrioventricular septum, from the atria to the ventricles.

9 Purkinje fibres.

10 It is myogenic. It can conduct electricity. It doesn't fatigue.

CONCEPT TEST

1 a During period A, the pressure in the atrium is higher than that in the ventricle. This means that blood will be moving from the atrium to the ventricle and, as a result, the atrioventricular valve will be open.

  b Atrial systole.

2 a The contraction of the muscular walls of the ventricle.

  b Blood isn't entering the ventricle, since ventricular pressure is higher than atrial pressure, but it isn't leaving the ventricle either, since the pressure in the ventricle is still lower than that in the aorta.

3 The 'dup' sound occurs when the semilunar valves close. This will happen when the pressure in the aorta first becomes higher than that in the ventricle and blood tries to run back into the ventricle from the aorta. On the graph, this occurs at about 0.4 seconds.

4 The pressure in the ventricle would not get as high. *Because the muscular walls of the right ventricle are not as thick as those of the left ventricle.*

UNIT 20

RECALL TEST

1 capillaries, venules, veins, the heart, arteries, arterioles.
2 In order to handle high-pressure blood without bursting.
3 They allow the exchange of substances between the blood and the surrounding cells.
4 To prevent low-pressure blood from flowing backwards.
5 Because a pressure peak, or pulse, is generated every time the heart contracts.
6 The watery part of the blood, excluding the cells.
7 4.
8 The concentration of oxygen.
9 The additional unloading of oxygen by haemoglobin which results from a high plasma CO_2 concentration.
10 The hollow space inside a blood vessel.

CONCEPT TEST
1

	carry oxygenated blood	connect to arterioles	involved in gas exchange	have a large lumen	show non-pulsed blood flow
Arteries	✔	✔	✗	✗	✗
Capillaries	✔	✔	✔	✗	✔
Veins	✗	✗	✗	✔	✔

2 Because their artery walls are less elastic, fluctuations in blood pressure will not be damped down as effectively and blood flow, even in the capillaries, may be uneven.

3 At high altitudes, where the llama lives, there is less oxygen in the air. To deal with this problem, the llama has evolved haemoglobin that has a higher affinity for oxygen than human haemoglobin. The llama haemoglobin picks up oxygen more readily at a low pO_2.

4 When there is a lot of CO_2 in the blood, this will combine with water to form a lot of carbonic acid. This carbonic acid will dissociate to produce a lot of hydrogen ions. These hydrogen ions will combine with oxyhaemoglobin and cause it to drop off oxygen; the Bohr effect.

UNIT 21

RECALL TEST

1 Transpiration is the movement of water through a plant. Translocation is the movement of organic molecules through a plant.
2 The evaporation of water from the leaves.
3 They have long, hair-like extensions which increase the surface area available for the absorption of water.
4 The apoplastic pathway is the movement of water via cell walls. The symplastic pathway is the movement of water from cell to cell and involves the crossing of cell walls and membranes.
5 Suberin.
6 The perforated end wall of a sieve-tube cell.
7 It contains mitochondria.
8 The source is the place where assimilates are produced. The sink is the place which they are transported to and where they are used up.
9 The active transport of organic molecules into the phloem, using energy.
10 Mass flow.

CONCEPT TEST

1 The fact that water pulls back from the cut point in both directions implies that the original water column, inside the stem, was stretched or under tension. This supports the cohesion–tension theory of transpiration, according to which the water column should be under tension. *Imagine stretching out a rubber band and then cutting it in the middle. Because the elastic band is stretched, it will spring away from the cut point in both directions, just like the water column does.*

2 a
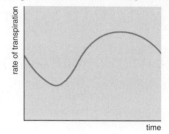

b The fact that the branch becomes narrower as the transpiration rate increases implies that transpiration is causing the sides of the xylem vessels, inside the branch, to be pulled together. This is what one would expect if the water in the xylem were under tension, as suggested by the cohesion–tension theory. *Water molecules not only attach to other water molecules, above and below them in the water column, but also to the sides of the xylem vessels. This means that if the water column is pulled up the xylem faster, at a higher transpiration rate, the water molecules will pull in on the sides of the xylem more and so the branch will become narrower.*

3

	contains lignified cells	contains cells with living contents	contains cells with incomplete end walls	transport is inhibited by metabolic poisons	transports the products of photosynthesis
Xylem	✔	✗	✔	✗	✗
Phloem	✗	✔	✔	✔	✔

UNIT 22

RECALL TEST

1 An organism that gains the organic molecules it needs by feeding on other organisms.
2 A heterotroph that feeds on a wide range of other organisms, including both plants and animals.
3 Pellagra.
4 *Any two from:* calcium, iron, phosphate.
5 It breaks food up into small pieces, which increases the surface area available for enzyme action.
6 It catalyses a reaction in which water is added, breaking a bond formed by condensation.
7 The process in which food is moved along the gut by waves of muscular contraction.
8 To prevent them from breaking down the cells in which they are formed.
9 Amylase. Trypsin. Lipase.
10 The reabsorption of water.

CONCEPT TEST

1 a Lipase splits lipid molecules, which are neutral, to produce glycerol and fatty acids, which are acidic, thus triggering a drop in pH.
  b Bile emulsifies the lipid molecules, increasing the surface area available for lipase to act on. This causes an increase in the rate of lipid breakdown and, as a result, a more rapid drop in pH.

2 If proteins are exposed to endopeptidases first, they will be broken up into many short pieces of polypeptide. This will increase the number of 'ends' available for exopeptidases to attach to, thus increasing the efficiency of protein breakdown.

3 **a** It has microvilli. These increase the surface area available for the absorption of materials.

    b It has mitochondria. These provide energy for the active absorption of materials.

UNIT 23

RECALL TEST

1 A zygote.

2 In one of the fallopian tubes.

3 This keeps the testes at a temperature lower than the rest of the body and ideal for sperm production.

4 The acrosome.

5 The post-reproductive phase of a woman's life.

6 A tiny cell that is produced and discarded after each division of meiosis, during egg production.

7 It allows implantation of the blastocyst and then provides it with nutrients.

8 The endometrium breaks down and is shed.

9 14 days.

10 It maintains the corpus luteum, so that this will carry on producing progesterone.

CONCEPT TEST

1 Aphids reproduce asexually during summer because the environment is friendly and because this is the quickest way to multiply. They reproduce sexually just before winter because this gives rise to offspring that vary, which means that at least some of them will have the right characteristics to survive over the harsh winter months.

2

Hormone	progesterone	LH	FSH	oestrogen
Site of production	produced by corpus luteum	produced by pituitary gland	produced by pituitary gland	produced by follicle
Effect	maintains uterine lining	triggers ovulation	triggers follicle development	causes uterine lining to develop

3 **a** The high levels of oestrogen and progesterone in the pill mimic the hormonal situation in pregnancy and prevent the menstrual cycle from starting up again. FSH production is inhibited, which means that the follicle fails to develop and no egg is released.

    b This causes the levels of oestrogen and progesterone to drop, temporarily, allowing the womb lining to be shed. The endometrium has to be shed, at regular intervals, to prevent infection.

UNIT 24

RECALL TEST

1 Angiosperms.

2 A flower which can produce both male and female gametes.

3 An anther and a filament.

4 8.

5 In the anthers.

6 In order to catch wind-blown pollen grains.

7 To control the growth of the pollen tube.

8 Because it involves two gamete fusion events; one between a male nucleus and the ovum and one between a male nucleus and the two polar nuclei.

9 The integuments.

10 A fruit.

CONCEPT TEST

1

	found in ovule	is the male gamete	produces a pollen tube	produced by meiosis and mitosis	stays in the flower
Embryosac	✔	✗	✗	✔	✔
Pollen grain	✗	✗	✔	✔	✗

2 **a** The tube nucleus controls the growth of the pollen tube. It is therefore active from the moment when the pollen tube starts growing, just after germination, to the point where it stops growing, just before fertilization.

    b There is a peak in the activity of the generative nucleus at about the time that the pollen tube stops growing. This activity results from the division of the generative nucleus to produce the two male nuclei that will be involved in double fertilization.

3 **a** A. *Gametes are haploid and carry only one allele for each characteristic. In this case, since the pollen grain had two A alleles, the generative nucleus has to end up with one A allele.*

    b Aa. *The zygote is formed from the fusion of a male nucleus, genotype A, and the ovum, or female gamete. The ovum has to be genotype a since it is a haploid product of the mother plant, which has the genotype aa.*

    c Aaa. *The endosperm food store develops from the triploid endosperm nucleus. This is formed by the fusion of a male nucleus, genotype A, and the two polar nuclei which, since they are identical to the ovum, must each have the genotype a.*

    d aa. *The fruit develops from the ovary which, since it is part of the original mother plant, must have the genotype aa.*

UNIT 25

RECALL TEST

1 Living organisms and the way that they interact with their environment.

2 *Any three from:* the air, oxygen, carbon dioxide, water, sunlight, etc. *You can list anything non-living here.*

3 A definable area containing a self-sustaining community of living creatures interacting with each other and with the abiotic environment.

4 Because it is not self-sustaining.

5 A group of organisms of the same species which live in a particular place at a particular time.

6 The way that the organism exploits, or makes a living from, its environment.

7 As sunlight.

8 Natural selection and the evolution of the species.

9 The relationship between nitrogen-fixing bacteria and the leguminous plants that they live inside.

10 Ecological succession.

CONCEPT TEST

1 **a** It consists of all the different populations of organisms that live inside the leaf; the populations of different microorganisms, the mosquito larvae population, the fly larvae population, etc.

    b

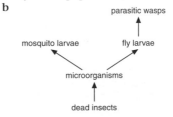

    c The leaf of the pitcher plant.

    d Pitcher plant leaf dwelling, microorganism eating. *i.e. they live in a pitcher plant leaf and feed on the microorganisms that they find there.*

    e Competition between mosquito larvae and fly larvae for the microorganisms that they both feed on.

    f Competition between different mosquito larvae for microorganisms. *Or any example of competition **within** a particular species.*

UNIT 26

RECALL TEST

1 Because they are the original source of organic molecules.
2 1%.
3 *One from:* Because some sunlight energy is reflected off the waxy cuticle of the leaf./Because some sunlight energy is transferred to the water inside the leaf, causing it to heat up and evaporate./Because some sunlight energy is of the wrong wavelength to be absorbed by chlorophyll.
4 net primary productivity = gross primary productivity – respiratory energy loss.
5 It is being eaten by decomposers.
6 As heat energy.
7 10%.
8 Because less energy is lost through egestion.
9 Because energy is lost between trophic levels and, eventually, there is not enough energy to support a further trophic level.
10 The total biomass of organisms at each trophic level.

CONCEPT TEST

1 a $\dfrac{10\ \text{MJ day}^{-1} - 3\ \text{MJ day}^{-1}}{10\ \text{MJ day}^{-1}} \times 100 = 70\%$

   The only component of the chicken feed that isn't absorbed across the chicken's gut is the component, with an energy content of 3 MJ day⁻¹, that is lost as faeces. The rest, with an energy content that works out at 7 MJ day⁻¹, is absorbed across the gut. Loss of energy due to respiration happens later, after absorption.

   b Battery chickens lose less energy through respiration, because they spend less energy moving around and generating heat to keep themselves warm. As a result, they convert a greater percentage of their chicken feed into biomass than free-range chickens do. *This is why it's cheaper to raise battery chickens than free range ones – unfortunately for the chickens.*

2 a Since energy is lost between trophic levels, there is less energy, and thus less biomass, at each successively higher trophic level. This is why the pyramid is pyramid shaped.
   b Because oak trees are very large and therefore, though they have the most biomass, all of this biomass is concentrated into a relatively small number of individuals.

UNIT 27

RECALL TEST

1 Nutrients are endlessly recycled. Energy just passes through the ecosystem, entering as sunlight and leaving as heat.
2 As hydrogencarbonate ions (HCO_3^-).
3 Photosynthesis.
4 Combustion.
5 Nitrate ions (NO_3^-).
6 The removal of amino groups from amino acids, producing ammonia (NH_3).
7 *Nitrosomonas* or *Nitrobacter*.
8 Because they use the energy obtained from chemical reactions to synthesise their own organic molecules.
9 Nitrogen gas (N_2).
10 The conversion of nitrogen gas (N_2) into ammonia (NH_3).

CONCEPT TEST

1 a The breakdown of the protein in the insect's bodies provides a source of nitrogen for insect-eating plants. They can use this nitrogen to manufacture their own proteins.
   b In a waterlogged soil, which is poorly aerated and lacks oxygen, denitrifying bacteria use up a lot of the available nitrates. Insect-eating provides an alternative source of nitrogen for plants growing in such nitrate-poor conditions.

2 a Leguminous plants contain nitrogen-fixing bacteria in their root nodules. These provide them with an additional source of nitrogen and allow them to out-compete normal plants on nitrate poor soil.
   b On nitrate-rich soil, leguminous plants are at a disadvantage when competing with normal plants because they have to support the nitrogen-fixing bacteria in their root nodules. This uses up energy which would be better invested in their own growth and yet gives them no advantage, because there is plenty of nitrogen available in the soil anyway.

UNIT 28

RECALL TEST

1 Because the animals that depended on the trees for food or for a place to live cannot survive without them.
2 The loss of nutrients from exposed soil.
3 The addition of artificial fertilizers.
4 Irrigation. The spraying of herbicides and pesticides.
5 When the products of fossil-fuel combustion mix with rainwater.
6 Because it contains dust particles and carbon monoxide from car exhausts.
7 *Any three from:* Carbon dioxide. Chlorofluorocarbons. Methane. Ozone.
8 A rise in global sea level and a disruption of global weather patterns.
9 Chlorofluorocarbons.
10 Because it protects the Earth from harmful ultraviolet radiation.

CONCEPT TEST

1 a Rainforest contains a greater amount of living plant biomass per m². This means that more photosynthesis will occur in a m² of rainforest and, as a result, more new plant material will be produced in each m² every year.
   b In rainforests, where the temperature is high, bacterial decomposition is very fast. As a result, even though a lot of plant litter is produced, it is broken down into inorganic molecules very quickly.
   c If a deciduous woodland is chopped down, there will still be a lot of organic material left in the soil, allowing rapid regrowth. In rainforest, deforestation will leave behind an organically poor soil that erodes easily and cannot support regrowth as effectively.

2 a The sewage, and the products of its decomposition, act as fertilizer, providing a source of inorganic ions that will speed up algal growth.
   b The large number of bacteria, which are feeding on the sewage and on decomposing algae, will use up all the oxygen in the water. This will kill off many freshwater animals.

INDEX

123